Transformation and sustainability in agriculture

Transformation and sustainability in agriculture

Connecting practice with social theory

edited by:
Sietze Vellema

Wageningen Academic
P u b l i s h e r s

Wageningen Academic Publishers
P.O. Box 220
6700 AE Wageningen
The Netherlands
www.WageningenAcademic.com
copyright@WageningenAcademic.com

ISBN 978-90-8686-161-3
e-ISBN: 978-90-8686-717-2
DOI: 10.3920/978-90-8686-717-2

First published, 2011

© **Wageningen Academic Publishers**
The Netherlands, 2011

Table of contents

Acknowledgements

The research, exchange and discussion that contributed to this volume was supported by the Netherlands Ministry of Agriculture, Nature and Food Quality through its Knowledge Base Funds (in the programme 'Transition Processes, Institutions, Governance and Policy', project KB-07-003-0001). I highly appreciate the sustained commitment of all the authors to this learning journey, which started in 2008 with a series of intense and creative working sessions bringing theory and practice together and which required open minds and respectful exchanges of ideas. The continued support of Krijn Poppe to make this project possible was of great value. The cross-boundary knowledge and actions of Jan Willem van der Schans were an inspiration in this process. Initially, Eric ten Pierick guided the research team through the pallet of transition literature. Michelle Luijben skilfully edited the contributions to this volume.

Wageningen, 2010

Sietze Vellema

Chapter 1

Transformation and sustainability in agriculture: connecting practice with social theory

Sietze Vellema

1.1 Context

This book is about how to analyse the social and institutional mechanisms that enable or hinder endeavours to change the way agriculture is done and food is provided. In recent decades, agriculture and food provision have faced a series of events often labelled as 'crises'. These have included damage due to increased pressure on natural environments, food scares related to contamination and animal diseases, public fears of genetic modification, and civil protests against the way animals are kept and treated. Since 2001, the Netherlands government has taken upon itself the vast task of initiating a selection of 'planned' transitions, in response to these persistent and complex problems and with the aim of achieving a sustainable society within a period of 30 years (Slingerland and Rabbinge, 2009). Sustainable agriculture is one of the selected transition areas, besides energy, mobility, and biodiversity and natural resources.

A transition arena provides room for long-term reflection and prolonged experimentation.[1] The transition arena for sustainable agriculture is an evolving network assembling a variety of stakeholders, projects and interventions in a joint endeavour to realise a transition of the agro-food complex. The authors contributing to this volume have, to varying degrees, been involved in projects, interventions and experiments embedded in this arena.

1.2 Management of transitions

Parallel to this combination of policy, practice and research, scholarly work has begun to theorise about transition processes, from the premise that such change processes do not have a predictable outcome. This literature adopts a strong process focus in describing the management of transitions and seeks to combine complexity theory with new governance models (Rotmans *et al.*, 2005; Loorbach and Rotmans, 2006). It does this with an open mind to participatory approaches for steering, to social learning and to iterative processes between shared vision and open-ended experimentation (Blanken *et al.*, 2004). This conceptualisation of transition processes also suggests the need for new forms of governance (Hendriks and

---

[1] The Dutch Research Institute for Transition (DRIFT) plays an important role in setting the knowledge agenda for transition thinking in the Netherlands. Terminology used in this volume is partly derived from its work (see http://www.drift.eur.nl/).

Grin, 2007). This literature brings on board concepts from a wide range of social science disciplines, such as sociology, culture sciences, and policy analysis. It has a strong interest in how small or incremental changes or technological improvements in strategic niches eventually result in a more drastic transformation of large systems (Geels and Raven, 2006). At the same time, it acknowledges that change processes towards sustainability comprise more than single technologies or organisations; they are the result of the combination and mutual reinforcement of technological, economic, ecological, social-cultural and institutional development at different levels (Rotmans *et al.*, 2001).

This volume uses the multi-level perspective (Geels and Schot, 2007) to visualise change processes, and its empirical contributions benefitted from this in picturing the chronology of events in specific change processes. Seeds of change can be observed in niches. Niche innovations can be self-invented or result from external pressures. The rules, routines and institutions driving continuity are located at the level of the socio-technical regime. Regimes show some level of stability, and from a transition perspective the question is what enables the reformation or reconfiguration of a regime. Next to the niche and regime level, the landscape level is considered to be largely exogenous to the transition process itself. But external pressures inducing change or creating windows of opportunity may come from this level, determined by slow changes in material infrastructure, culture, social structures, worldviews and paradigms.

Consequently, conceptualisation of a transition process shifts attention to the level of socio-technical systems (Geels, 2004). It sees transition processes not as the result of identifiable actors or factors, but caused by failures of a systemic nature (Kemp *et al.*, 2007). This is reflected in the presentation of transition processes in terms of speed and direction, and also as an S-curve revealing the seed of change gradually reconfiguring and stabilising a new socio-technical regime (Loorbach and Rotmans, 2006). This visualisation helps us to consider the long-term and systemic implications of change processes, which, in the short term, may look whimsical, and to work on intermittent and partial solutions for a systemic problem (Rotmans *et al.*, 2005).

1.3 Transition management and theoretical pluralism

Transition management seems to have a strong interest in building systems, and it strongly emphasises technological experimentation (Hegger *et al.*, 2007). This may result in a neglect of various dimensions of human behaviour and social organisation; and it may ignore experiments with forms of social organisation, institutional arrangements and behavioural patterns (for an example see Vellema *et al.*, 2006). The premise of the work presented in this volume is that to understand what enables or constrains processes of change it is necessary to further develop a socialised perspective on the management of transitions and to theorise more explicitly about how conditions at the regime level, manifested in rules, routines and behavioural patterns, both shape the conditions for scaling up micro experiments and respond to signals from these experiments.

The literature on transition management under conditions of uncertainty and complexity is explicitly interested in the capacity to (re)engineer systems, the capacity to steer processes of socio-technical change, and the capacity to scale up experiments as a pathway to modify or transform regimes. The contribution of this volume is to further unpack these capacities by mirroring transition management thinking in a selection of social theories. In contrast to what transition management tries to do, namely attempting to integrate everything into one framework, this volume explores an approach that rests on a multiplicity of competing theories, as proposed by Dewulf *et al.* (2009). Hence, the volume comprises six theoretical essays, each presenting the merits of an individual theory. Each of these distinct theories is then applied in four case studies of change processes in agriculture and food provision. This is unlike a process which seeks to integrate distinct theoretical accounts and empirical processes into a single framework.

This volume reflects an endeavour to encourage theoretical pluralism. The project in which this volume was realised invited a number of theoretically interested researchers at or related to Wageningen University and Research Centre, to produce a theoretical essay for further use in a dialogue with more practice-oriented researchers. The essays were discussed in the team, and the usefulness of the theoretical insights was concretised in the context of four case studies. Towards the end of the joint exercise, the so-called 'theorists' were challenged to elaborate on the action perspectives or intervention repertoires hidden within the theoretical exposes. This served to further link the theories with the reality of practitioners, policy-makers and transition managers, who must take informed decisions about how to act and make change processes work within a specific context.

The social theories in this volume were selected along two axes, to ensure variety, and with a specific interest in theories that focus attention on the level of the socio-technical regime. Theories were placed on the 'actor-system' axis and on the 'idealism-materialism' axis. Gerdien Meijerink examines how new institutional economics, as developed by Douglass North and Masahiko Aoki (Chapter 2), combines a focus on individual actors with incentives for changing behaviour. A system perspective on human behaviour and social engineering is reflected in the work of Niklas Luhmann, presented in the chapter by Kristof van Assche, Martijn Duineveld, Gert Verschraegen, Roel During and Raoul Beunen (Chapter 3). The work of Mark Granovetter, discussed by Machiel Reinders (Chapter 4), relates change to both a micro perspective on individual behaviour and a meso perspective on patterns of interaction within social networks. Trond Selnes and Catrien Termeer (Chapter 5) use the work of Karl Weick to examine how actors construct and make sense of change under conditions of uncertainty within organisations. Martijn Duineveld and Guus Dix (Chapter 6) draw on the work of Michel Foucault to detect three mechanisms – disciplining, subjection and exclusion – which make explicit how the entanglement of power and knowledge conditions human behaviour and social action. In Chapter 7, Sietze Vellema builds on the work of Ted Benton to incorporate non-manipulable material and natural limits on human-driven change into the analysis of transitions of socio-technical systems.

1.4 Connecting theory and practice

Combining theories provides grounds for a discussion on institutional failures in and the human dimension of transition processes. This is done using four case studies. Moreover, from the practitioners perspective it seems relevant to discern how actors, i.e. the partners in a transition process, behave under certain institutional and material conditions and to identify their room for manoeuvre and their capacity to change matters. Transition is also about making choices and agreeing to cooperate to make things work differently. Therefore, it is relevant to incorporate the mechanisms that induce choices and that install institutional arrangements into understanding systemic change, rather than merely projecting systemic change as an outcome of bounded, mainly technical inventions and experiments that set the socio-technical system in motion in a new direction. The aim in this volume is to find a way to conceptualise transition as an evolving configuration of social and technical realms, of which the precise outcomes are contingent on social choice and behaviour as well as on technological and natural realities.

The four case studies centre on concrete change processes which are embedded in a possible larger transition towards sustainable agriculture and food provision. Each case study reconstructs the chronology of events constituting the change process, and then tries to capture this in terms of transition management. Next, each draws on the six theoretical essays to explore what one can see or expect to observe in the case study when using the set of lenses provided by each distinct theory. The value of this exercise is to allow us to look deeper into the process, and into the mechanisms that bring about change.

In the first case study, Jan Buurma and Marc Ruijs (Chapter 8) seek an explanation of the different development pathways of two horticultural expansion areas in the northern and eastern Netherlands. Both transition processes were driven by national policies and plans to reform agricultural production and to change the way agricultural production is embedded in a specific environment. Differences in the development pace and outcome of the two areas seem to be related to the behaviour of actors and the relations between actors. A similar insight arises from the second case study, in which Carolien de Lauwere and Sietze Vellema (Chapter 9) examine how endeavours to reconstruct livestock farming, partly induced by outbreaks of animal diseases and concerns about animal welfare, touch ground in specific social environments. The case study seeks explanations for local resistance to a potential change towards a more environmentally benign and animal friendly system of livestock farming. In so doing, it is also concerned with how generic and imposed transitions are articulated in the logics embedded in specific localities.

The third case study, by Rolien Wiersinga, Derek Eaton and Myrtille Danse (Chapter 10), focuses on a business model for supplying smallholder vegetable farmers in the South with improved planting materials. The business model is embedded in processes of commercialisation and specialisation in rural areas that have a long history of producing agricultural crops mainly for food security. From the case study, it becomes clear that this change process, linked to general strategies for reducing poverty, is a combination of existing proven technological principles (i.e. hybridisation of seed) and a new institutional architecture featuring a division of labour between farmers and seed companies. Also, the business model is accompanied by new cultural models of what farmers are supposed to be, e.g. entrepreneurs operating in a market-driven environment. The application of a pallet of theoretical insights to the case study opens the black box of what seems to be a straightforward business model and creates space for counterfactual reasoning.

In the last case study, Jan Buurma (Chapter 11) shows that the court proceedings initiated by environmental groups, in the context of fierce public debate, was an important trigger for transforming the pest control regime in Dutch agriculture. It confirms that paying attention to social or institutional experiments is as important as framing transitions in terms of technological novelties. As a result of polarisation and the use of the juridical system, new alliances were brokered which set the way for a drastic modification of the regime of crop protection/pesticide application in the Netherlands. Here again, employing the distinct theories helps us to unravel various dimensions linked to this process of institutional change.

1.5 A word on methodological choice

This volume's collection of theoretical essays and case studies opens a discussion on connecting practice with social theory. This required a methodological endeavour to enable theoretical pluralism. The chosen approach enriched the reconstruction of the selected case studies by encouraging a theory-informed search for mechanisms hidden within the change process (Gerring, 2007). A transition perspective helped us to look at outcomes of change processes from a long-term and systemic viewpoint. The contribution of the work presented here is to enable us to open the black box of time and to place specific change processes by focusing attention on the mechanisms, with causal properties, that produce institutional change. This provides the basis for theory-driven empirical inquiry and leads to a research design based on the combination of researchable, alternative hypotheses. Methodologically, it suggests confederating empirical inquiry with a network of theories. This will make it possible to borrow, consolidate, or pass on explanatory structures and to relate a set of statements or hypotheses to segregated observations. This volume takes a first step in doing so.

Theoretical pluralism, i.e. pitting alternative theories against the same body of data (Denzin (1970) uses the term 'theoretical triangulation'), may be able to achieve the blend between theory and empirical research necessary to get inside the black box of transitions. Since transitions occur usually once, and do not repeat themselves in precisely the same way, it is important to find methodological handles to detect traces of the causal powers at work in specific cases (Blatter and Blume, 2008). This volume suggests that future analyses of transition processes need to find a way to incorporate theory-driven expectations and to allow competing candidate mechanisms to be assessed in relation to data associated with the mechanisms themselves and with the context and outcomes concerned (Perri 6, peronal communication, 2008). The often whimsical and uncertain reality of practice may benefit from theoretical pluralism, as it permits new and unexpected findings to challenge the assumptions prevalent in policy- and planning-driven transition processes.

References

Blanken, H., A. Loeber and D.J. Joustra (2004) Veranderen is leren, leren is veranderen. NIDO Transitiepaper 9, The Hague: Dutch National Initiative for Sustainable Development (NIDO) and Ministry of Housing, Spatial Planning and the Environment (VROM).

Blatter, J. and T. Blume (2008) In search of co-variance, causal mechanisms or congruence? Towards a plural understanding of case studies. Swiss Political Science Review 14 (2): 315-56.

Denzin, N.K. (1970) The Research Act: a theoretical introduction to sociological methods. Aldine, Chicago, IL, USA.

Dewulf, A.E., C.J.A.M. Termeer, R.A. Werkman, G.R.P.J. Breeman and K.J. Poppe (2009) Transition management for sustainability: towards a multiple theory approach. In: K.J. Poppe, C. Termeer and M. Slingerland (eds.). Transitions Towards Sustainable Agriculture and Food Chains in Peri-Urban Areas. Wageningen Academic Publishers, Wageningen, the Netherlands, pp. 25-50.

Geels, F. (2004) From sectoral systems of innovation to socio-technical systems: insights about dynamics and change from sociology and institutional theory. Research Policy 33 (6-7): 897-920.

Geels, F. and R. Raven (2006) Non-linearity and expectations in niche-development trajectories: ups and downs in Dutch biogas development (1973-2003). Technology Analysis and Strategic Management 18 (3/4): 375-92.

Geels, F. and J.W. Schot (2007) Typology of sociotechnical transition pathways. Research Policy 36 (3): 399-417.

Gerring, J. (2007) The mechanismic worldview: thinking inside the box. British Journal of Political Science 38 (1): 161-79.

Hegger, D.L.T., J. Van Vliet and B.J.M. Van Vliet (2007) Niche management and its contribution to regime change: the case of innovation in sanitation. Technology Analysis and Strategic Management 19 (6): 729-46.

Hendriks, C.M. and J. Grin (2007) Contextualizing reflexive governance: the politics of Dutch transitions to sustainability. Journal of Environmental Policy and Planning 9 (3-4): 333-50.

Kemp, R., D. Loorbach and J. Rotmans (2007) Transition management as a model for managing processes of co-evolution towards sustainable development. International Journal of Sustainable Development and World Ecology 14 (1): 78-91.

Loorbach, D. and J. Rotmans (2006) Managing transitions for sustainable development. In: X. Wieczorek and A.J. Wieczorek (eds.). Understanding Industrial Transformation: views from different disciplines. Springer, Dordrecht, the Netherlands.

Rotmans, J., R. Kemp and M.B.A.V. Asselt (2001) More evolution than revolution: transition management in public policy. Foresight 3 (1): 15-32.

Rotmans, J., D. Loorbach and R. Van der Brugge (2005) Transitiemanagement en duurzame ontwikkeling; co-evolutionaire sturing in het licht van complexiteit. Beleidswetenschap 19 (2): 3-23.

Slingerland, M. and R. Rabbinge (2009) Introduction. In: K.J. Poppe, C. Termeer and M. Slingerland (eds.). Transitions Towards Sustainable Agriculture and Food Chains in Peri-Urban Areas. Wageningen Academic Publishers, Wageningen, the Netherlands, pp. 13-23.

Vellema, S., D. Loorbach and P. Van Notten (2006) Strategic transparency between food chain and society: cultural perspective images on the future of farmed salmon. Production Planning and Control 17 (6): 624-32.

Chapter 2

New institutional economics: Douglass North and Masahiko Aoki

Gerdien Meijerink

2.1 Introduction

This chapter discusses the possible contribution of new institutional economics (NIE) to the transition literature. The two fields seem to have developed in parallel universes, and there has been no cross-fertilisation. This essay presents some possible starting points for exchange. In an overview of NIE, Brousseau and Glachant (2002: 4) say, '[T]he strength of NIE lies in its proposal to analyze governance and coordination in all sets of social arrangements.' They warn, however, that 'the design of institutional systems is not based on optimisation computation but on trial and error, on the implementation of solutions that should be recognized as imperfect and temporary…. In such a context, it is essential to take into account the management of changes together with the processes of evolution.' NIE thus has something to say about managing transitions, although it might provide a slightly different perspective on change management than that in the transition literature.

This chapter first discusses how NIE explains social change, including a review of some concepts used in NIE. It then shifts focus to what NIE adds to the literature on transition management and looks at tensions between NIE and some of the assumptions of transition management. It concludes by providing starting points for using concepts of NIE in management of change. The chapter draws primarily on two authors from the NIE school of thinking: Douglass North (1990, 2005) and Masahiko Aoki (2001, 2007). North developed several critical ideas about what institutions are, and Aoki elaborated on these. Other literature is used to expand on or back up their thinking.

2.2 NIE's explanation of social change

2.2.1 Institutions

The well-known definition by North (1990) of an institution is that it consists of the rules of the game in a society. Aoki (2001, 2007) elaborates on this and proposes that institutions are collectively recognised rules and symbols plus behavioural beliefs (expectations) of agents about other players' choices and intentions. Players base their own behaviour (strategies, actions, etc.) on these beliefs. These sets of beliefs are also called 'mental models' (Denzau and North, 1994) or 'social models' (Eggertsson, 2005). 'Meaningful rules' are those that are generally respected and obeyed. For instance, if a rule says to stop at red traffic lights (a symbol), and those around you appear to be doing so, you will probably do so too. But if no one seems to be obeying, you will probably stop obeying the rule too. Thus the rule 'stop at red traffic lights' ceases to be meaningful.

Aoki (2007) emphasises that institutions are self-sustaining, salient patterns of social interactions. The rule that says to drive in the right-hand lane is self-sustaining ('if everyone else does it, I will too, though I don't need to know the intricate traffic laws behind it'). The rule is also salient ('if I don't drive in the right-hand lane, I'll crash into other cars'). Moreover, it represents a pattern of social interaction ('I see others doing it, so I join in, and others do the same'). The rule to drive in the right-hand lane is also meaningful, as it makes sense and is generally respected. Everyone who drives knows it and can be confident that other drivers know it too.

Salience is important, because human beings have limited capacity for storing and processing information. Or, according to Greif and Laitin (2004: 638), 'The information compressed in socially transmitted rules enables individuals without knowledge of all the relevant parameters and causal mechanisms, and with limited computational ability, to act in a manner that leads to equilibrium behaviour.' Thus, institutions 'aggregate, in equilibrium, the dispersed information that each of these individuals has. In other words, these rules both provide individuals with the information they need to make decisions regarding how to act as well as aggregating the information privately held by each of them' (Greif and Laitin, 2004: 638). Although traffic laws are written, individuals participating in traffic are not expected to know these laws by heart. It is sufficient that they know the most important rules, which they either learn in a formal way (i.e. passing an exam before obtaining a driving licence) or informally (by observing the behaviour of others). According to Greif and Laitin, this also explains why institutions remain relatively stable.

Greif and Laitin (2004) add two other reasons why salience is important: limited attention and coordination. Regarding the first, people's attention is a scarce resource. When we drive in busy traffic, we cannot absorb all of the information out there. Here institutionalised rules come to the rescue. They enable us to choose appropriate behaviour in complicated situations while devoting limited attention to decision making. For instance, when the traffic light is green we do not need to monitor all the cars in cross-wise traffic, because we can be confident that drivers wanting to cross the intersection will wait.

Coordination is the second reason why salience is important, as discussed by Greif and Laitin. By this we mean how multiple actors with different interests and information can coordinate their behaviour. This problem is especially acute when a situation marginally changes:

> [I]ndividuals face the problem of which behaviour to follow in the new situation, given the multiplicity of self-enforcing behaviours. Because people do not share the expectations that some new equilibrium behaviour will be followed, they

are likely to rely on past rules of behaviour to guide them and to continue following past patterns of self-enforcing behaviour. There are many reasons why such coordination may fail to transpire even when it is beneficial. Sunk costs associated with coordinating change, free-rider problems, distributional issues, uncertainties, limited understanding of alternatives, and asymmetric information may hinder coordination on new behaviour. (Greif and Laitin, 2004: 638)

For instance, when the white lines dividing the road disappear, people will tend to stick to the rules, such as 'drive on the right-hand side'.

Implementing new rules, such as switching to driving on the left side of the road, is extremely difficult; not just because of the sunk costs involved (e.g. past investments in road signs and cars with the steering wheel on the left side) but also because of people's difficulty in learning new habits (driving on the left).

2.2.2 Framework

Institutions, defined as 'rules cum shared beliefs', can be depicted as in the framework by Aoki (2001) (Figure 2.1). Institutions are represented by the lower boxes in the figure. The 'play of the game' refers to the strategies (actual behaviours) of individuals, and these are based on their beliefs. It may constitute an equilibrium, as in the case of driving in the right-hand

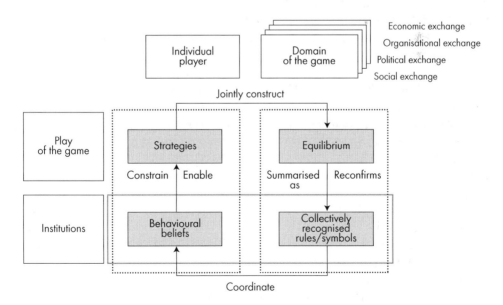

Figure 2.1. Institutions as rules cum beliefs (based on Aoki, 2001).

lane. This equilibrium then reconfirms the collectively recognised rules (everyone knows to drive in the right lane). But it also reconfirms itself in collectively recognised rules and symbols. I need only to see a symbol such as a signpost to know that the rule is indeed to drive on the right (while in the UK I quickly learn that it is the other way around). Or I need only to see several other people driving in the right lane (or see them in the left lane in the UK). However, Aoki adds that 'equilibrium' does not mean that multiple equilibria are not possible. Institutions are humanly devised constructs and may be diverse. Thus, for instance, there may be local differences in how things are done.

A key feature of NIE (and economics in general), but which is missing from the figure, is enforcement, which Aoki (2001: 8) integrates into the strategies and behavioural beliefs boxes. If someone acts in such and such a way, then some other actor (say, a court or the government) will typically act in such and such a way (so that such and such a consequence will befall him/her). Stopping at red traffic lights is enforced by the police, making it an equilibrium institution. But there may also be informal ways to enforce rules, such as the outrage displayed by people on the street if a driver does not stop. Thus, behavioural beliefs and strategies include feedback mechanisms.

Traffic rules such as 'stop when the traffic light is red' or 'drive in the right-hand lane' are very simple illustrations. Yet Aoki recognises that actual institutional dynamics are likely to also involve interactions of economic, organisational, political and social factors. These complicate matters enormously and are captured by the 'domain of the game' in Figure 2.1. Aoki suggests that a first step towards the analytical treatment of this interactive process is to specify prototypes of domains that may capture some minimally essential elements of each factor that may arise prior to interactions. He distinguishes four prototypes:

- *The economic exchange domain.* This is the domain in which transactions of private goods take place....
- *The organisational exchange domain and organisational field.* The organisation may be a player of the game in an economic exchange domain. At the same time, the organisation itself may be regarded as emerging as an institution in the domain of work collaboration....
- *The political exchange domain....* [T]his domain in its prototype is composed of two types of agents: the government and multiple private agents. This asymmetric structure is somewhat similar to that of the prototype organisation emerging in the collaborative work domain. They are different, however, in that...in the political exchange domain the exit option is not open to the private agents. The government can provide public goods to the private agents...in exchange for the extraction of costs in the form of taxes, issuing of money, etc. But the fact that the government has such power may also imply that it may have power to transgress the various rights of the private agents.... The private agents may respond by supporting/resisting/submitting-to the government's choice...with/without mutual coordination among themselves.
- *The social exchange domain.* This domain may be conceptualized as the one in which social symbols (languages, rituals, gestures, gifts, etc.) that directly affect the payoffs of

players, such as esteem, emotional rejection, sympathy, benign neglect, and so on are unilaterally delivered and/or exchanged with 'unspecified obligations to reciprocate'.... Institutions that arise in this type of domain are identifiable with social customs and norms enforced by the threat of social ostracism. (Aoki, 2007: 13-16)

2.2.3 Institutional change

To analyse institutional change, the starting point is again the individual:

> [W]hen an existing set of rules does not produce satisfactory results relative to an agent's aspirations, the agents may start questioning the relevance as well as usefulness of their own subjective game models. In particular, they may search for and experiment on new strategic choices (rules) involving the expansion of the repertory of action choices. (Aoki, 2001: 239-240)

Aoki labels such a gap between aspiration and achievement 'general cognitive disequilibrium' when it is experienced by a critical mass of individuals. This can happen, for instance, when there are technological innovations that make new action choices feasible, or when external shocks occur, such as defeat in war. These are exogenous (environmental events). Internal cumulative consequences of certain rules can generate a change in the distribution of assets, power and expected roles among agents, which may lead to problems in the enforceability of those rules. For instance, groups of individuals may come to consider rules as unfair and start resisting them.

North (1993) calls individuals who induce change 'entrepreneurs'. Aoki (2001) explains that agents may appear who start to re-examine the effectiveness of their own activated choice sets and discover novel actions or a new Schumpeterian bundling of domains, thus expanding the set of strategic choices open to them. Successful new choices will likely be imitated by other agents.

Aoki explains that as the

> implementation of new choices begins, the existing institution will cease to provide a useful guide for individual choices. It will be incapable of providing an effective summary representation of newly emergent choice profiles and thus cannot be helpful in informing agents' expectations. This is what is meant by an institutional crisis. The taken-for-grantedness of the old institution [is] called into question. (Aoki, 2001: 241).

Agents have to process more information and form expectations regarding emergent patterns of choices by others that may be relevant to their payoffs. Aoki (2001: 241) describes what may happen in such an institutional crisis: agents may try to follow practices they see operating effectively in other domains (e.g. in other countries). In polity domains there may be a few alternative discourses or ideas that compete with each other and may help in designing a new policy (Aoki, 2001: 241):

> *A political leader or entrepreneur may try to signal a desired direction of change by a symbolic action.... [Or,] the sensational public disclosure of some untoward behaviour that was tolerated under the normal state may have a decisive impact on agents' perception of what is or is not a proper choice of action. So a few systems of predictive and normative beliefs should emerge and compete with each other. Competition among these beliefs characterizes the transitional process. Which competing system becomes a focal point where the expectations among agents converge, and thus [becomes] a candidate for a new institution will depend on how learning, emulation, adaptation, and inertia interact across economic, political and social exchange domains and become stabilized. (Aoki, 2001: 241-2)*

This process is summarised in Figure 2.2.

In this context, Eggertsson (2005) describes the possibilities and limits of reform in his book on imperfect institutions. Institutional change, he says, does not always lead to more efficient institutions and to economic growth. North (1990) underlines the importance of path dependence. In economic theory, he argues, there has been a tendency to think that over time, inefficient institutions are 'weeded out'. Efficient ones are said to survive, which leads

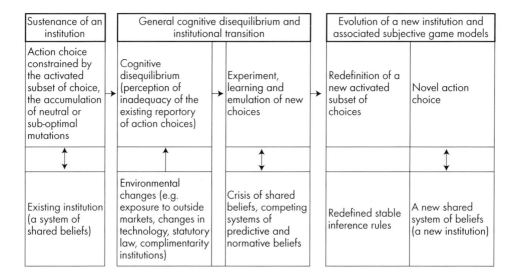

Figure 2.2. Mechanism of institutional change (based on Aoki, 2001).

to the assumption of 'a gradual evolution of more efficient forms of economic, political and social organisations' (North, 1990: 92). However, he rhetorically asks:

But why would relatively inefficient economies persist? What prevents them from adopting the institutions of the more efficient economies? If institutions existed in the zero transaction cost framework, then history would not matter; a change in relative prices of preferences would induce an immediate restructuring of institutions to adjust efficiently.... But if the process by which we arrive at today's institutions is relevant and constrains future choices, then not only does history matter but persistent poor performance and long-run divergent patterns of development stem from a common source. (North, 1990: 93)

North is referring here to institutional path dependence, and not technical path dependence, which is also frequently discussed. Technical path dependence is often illustrated by the example of the position of letters on a keyboard (QWERTY). This might not be the most efficient, but once people have learned it, the cost of learning a new set-up is prohibitively high. Institutional self-reinforcement arises from several mechanisms:
- large initial set-up costs when the institutions are created;
- significant organisational learning effects that arise from the opportunity set provided by the institutional framework;
- 'lock-in', meaning the difficulty of exit once a solution is reached;
- 'path dependence' or the sequence of minor events and changes that determine the particular path of solutions (Arthur, 1988, cited in North, 1990).

2.3 Applying NIE to the field of transition

The selected NIE literature has much in common with the transition literature, despite differences in the language in which their concepts are framed. The work by Geels and co-authors is in places strikingly similar to that of Aoki described above. A figure by Geels depicting actor-rule system dynamics resembles that of Aoki, but the different theoretical grounding leads to a different structure and consistency (Figure 2.3).

Both the NIE and transition literature focus on the importance of belief systems and the related behaviour of human actors and on the fact that transition involves a change in those belief systems and thus in behaviour. The transition literature zooms in on projects, niches, stakeholder platforms, transition arenas, pilots and discursive spheres. Herein, a limited group of actors (stakeholders) experiment, innovate and exchange information, ideas and opinions, debating and steering towards change. Transition authors underline the fact that by no means does this process imply a blueprint or a linear process. Rather, the process is reflexive or iterative, in which the precise direction or outcome is not known in advance.

In this respect, the transition literature is often in line with evolutionary economics and NIE. In the simplest of terms, the main difference between evolutionary economics and NIE is that

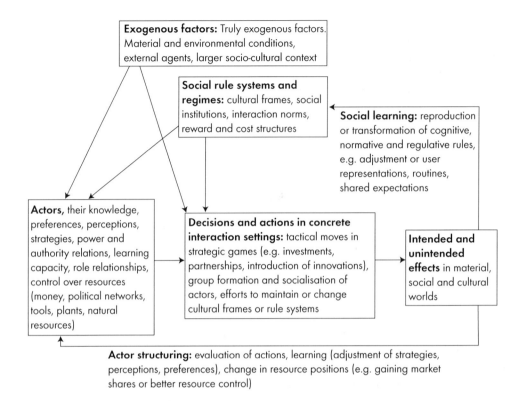

Figure 2.3. Geels' depiction of actor-rule system dynamics (Geels, 2004: 908).

the first focuses on new technologies and new organisational set-ups and their (commercial) success, while the latter focuses solely on institutions. There is much overlap however (see Coriat and Dosi (1998) for an overview of evolutionary economics versus NIE).

There are also some clear differences between the transition literature and that on NIE. I highlight three: (1) enforcement and incentives, (2) (bargaining) power and (3) the importance of the cognitive sciences. These are central themes in NIE, but are not (explicitly) taken into account in the transition literature. Many others can be mentioned as well, such as the importance of property rights and transaction costs. However, these are not really critical to the transition literature, or they are incorporated in the three issues focused on here.

2.3.1 Enforcement and incentives

Enforcement in general is a key concept in economics, especially in NIE. Aoki makes clear that if rules are not followed, they are not relevant. NIE authors therefore devote significant attention to why people do follow certain rules. Two reasons emerge as important: enforcement and incentives.

Enforcement can be conceptualised in two broad ways. One is official enforcement in a society through established rules, laws and other means to ensure that people behave according to the rules. Another is informal enforcement (for lack of a better term), which basically consists of norms, values and social sanctions of people in a society towards those who do not comply. It is important to realise that the rules and norms are only institutions when they are expressed in behaviour.

Incentives can be seen as the flipside of enforcement. When a group of people have an interest or stake ('a payoff' in the language of Aoki) in adopting certain behaviour, the institution will be self-sustaining. This also has implications for implementing (institutional) change. One can compel people to change their behaviour through enforcement. However, incentives can motivate people to change as well. '[E]conomies of scope, complementarities, and network externalities of [institutions]…make institutional change overwhelmingly incremental and path dependent' (North, 2005: 59).

2.3.2 Bargaining power

The transition literature focuses almost uniquely on experimentation and communication (learning, visions, knowledge and information sharing), and seems to assume that both will lead to (radical) change. In NIE, learning plays a central role as well (in changing belief systems), and NIE is increasingly linked with the cognitive sciences. However, NIE also stresses (bargaining) power and vested interests as important factors for achieving change, and these are not explicitly addressed in the transition literature.

Loorbach and Rotmans (2006: 9), for instance, look at 'networks of innovators and visionaries that develop long-term visions and images that, in turn, are the basis for the development of transition-agendas and transition-experiments, involving growing numbers of actors'. Geels and Raven (2006) restrict their focus to learning, expectations and visions about innovations. These are to be realised through experimental, pilot or demonstration projects that 'provide space for interactions between actors and the building of social networks…[and can be diffused and] compete with existing technological regimes that are characterised by lock-in and path dependence' (Geels and Raven, 2006: 376). Geels (2004: 898) stresses the learning aspects of integrating new technologies by users.

North (2005: 216) contrasts this view by stating that 'it is the bargaining strength of the individuals and organisations that counts. Hence only when it is the interest of those with sufficient bargaining strength to alter the formal rules, will there be major changes in the formal institutional framework'. Unwittingly, Hendriks and Grin (2007) describe a case study that actually clarifies the shift of bargaining power from the 'iron triangle' to 'outsiders'. They explicitly point out that this shift involves a renegotiation of who may make decisions, when and on what basis. However, bargaining power, interests (and enforcement) do not feature in their conceptual framework. They seem to emphasise changing opinions and convictions.

2.3.3 Link with the cognitive sciences

NIE underlines the importance of belief systems (mental models), ideologies, learning and knowledge (see also the work of Friedrich Hayek and other economists from the Austrian School). A key insight of NIE, and particularly Aoki, is that institutions consist of rules cum beliefs, which opens up a whole new area of research integrating NIE literature with that of the cognitive sciences:

> *The way we perceive the world and construct our explanations about that world requires that we delve into how the mind and brain work – the subject matter of cognitive science. This field is still in its infancy but already enough progress has been made to suggest important implications for exploring social phenomena. Issues include how humans respond to uncertainty and particularly the uncertainty arising from the changing human landscape, the nature of human learning, the relationship between human learning and beliefs systems, and the implications of consciousness and human intentionality for the structure that humans impose on their environment. (North, 2005: 5)*

Hendriks and Grin (2007) is interesting in this respect because it analyses communication, the (formal or informal) diffusion of ideas. This dovetails with the concept of belief systems. Also reflexivity ties in with the ideas in NIE on learning and feedback (through other people's behaviour). However, the link with the cognitive sciences seems to be absent in the transition literature.

2.4 Tensions between NIE and the assumptions of transition management

Perhaps the biggest analytical differences between NIE and transition management can be found in how change is conceptualised and appreciated. The transition management literature focuses on 'how to change things' while NIE focuses on 'how things change'. Transition management is normative, while much of the new NIE literature is not. The main judgement that NIE makes about institutions is whether or not they contribute to economic growth.

The transition literature basically assumes that through (experimental) action (within projects), change towards a (preconceived) goal can be achieved. Thus, many of the figures and graphs from this field show arrows pointing upwards, implicitly assuming that the new situation is an improved one. The transition literature is replete with such diagrams. NIE authors seldom use figures with arrows pointing in a certain direction (with the exception perhaps of Aoki, who uses them very sparsely). The new NIE literature does not tend to make such assumptions. It also examines change that does not necessarily constitute an improvement. Recently, for instance, North *et al.* (2009) describe why institutional change leading to economic development is so difficult to attain in many developing countries. They view developed countries, in which the formation of a set of institutions has led to democracy and economic growth, as the exception, not as the norm.

Loorbach and Rotmans (2006) are relatively careful about capacity to manage change. According to them, 'transitions cannot be managed in the traditional sense [because] transitions are the result of the interplay of many unlike processes, several of which are beyond the scope of management' (Loorbach and Rotmans, 2006: 8). They do believe that management can 'influence the direction and speed of a transition through various types of steering and coordination' (Loorbach and Rotmans, 2006: 9). North suggests that incremental changes at particular margins are possible. These 'small changes in both formal rules and informal constraints will gradually alter the institutional framework over time, so that it evolves into a different set of choices than it began with' (1990: 68). Here, North's so-called 'entrepreneurs' can play an important role: '[I]ncremental changes come from the perception of the entrepreneurs in political and economic organisations that they could do better by altering the existing institutional framework at some margin' (North, 1990: 8). However, he hastens to point out that these entrepreneurs 'frequently must act on incomplete information and process the information that they do receive through mental constructs that can result in persistently inefficient paths' (North, 1990: 8). Similarly, Aoki (2001: 16) states that although a multiplicity of equilibria are possible and although institutions are humanly devised, 'they can be neither arbitrarily designed nor discretionary implemented'.

Like North, Hendriks and Grin (2007) conclude that change is incremental and at the margins, though they seem slightly surprised and maybe even disappointed by this. 'The final insight to highlight from this case study [the Gideon project] is that the impact of reflexive arrangements can be subtle and indirect. They rarely culminate in grand policy decisions made by formal institutions, such as relevant ministries or even the Parliament' (Hendriks and Grin, 2007: 345).

2.5 Perspectives for acting

As mentioned above, NIE says very little about the goal of transition processes concerning contents. As an economic discipline, one of the implicit goals is economic development. Although North's 2005 book is entitled *Understanding the Process of Economic Change* (and not growth!) the main question posed by North and colleagues is: 'what historical process(es) have generated institutions in a handful of countries capable of sustained economic development in the twentieth century, while most countries still fail to develop thriving markets, competitive and stable politics, and cultures that promote deep human capital accumulation for most of their populations?' (North *et al.*, 2009: 1). The Dutch transition literature seems to take for granted the institutions that have generated sustained economic development, and focuses on what are basically unwanted effects (externalities), such as environmental deterioration, which have not (yet) been corrected by the institutions.

The question remains as to what perspectives to act does the NIE literature offer. In other words, what intervention theory does NIE embody?[2] First and foremost, NIE can inform social planners by providing a better understanding of how institutional frameworks produce economic outcomes. NIE does this by analysing how individuals and groups play with and bypass institutions. In an overview of NIE, Brousseau and Glachant (2002: 19) call this pragmatism, which is a response to the ability of agents to behave strategically and bypass collective constraints to protect their interests and enhance their individual wealth. Often (incremental) institutional change is achieved by accumulation and propagation of micro-institutional reforms for fixing local problems.

2.6 Conclusion

Managing change has never really been the focus of NIE. The closest that NIE comes to this topic is in the overlap with development economics, where the complexity of managing institutional change is highlighted. However, the NIE literature demonstrates that changing or manipulating beliefs is essential since merely changing formal rules is insufficient to impact actual business and social practices. Perhaps more importantly it emphasises implementing institutional patches, which consist of light and local reforms to procedures (e.g. ad hoc licenses, decrees and administrative procedures). An attempt to directly change, for instance, the distribution of property rights would probably lead to clashes among interest groups, with incumbents able to protect their rights by their veto powers. Brousseau and Glachant (2002: 21) therefore suggest the gradual implementation of institutional patches that are aimed at fixing the worst effects of institutional regimes. This will not guarantee success, but it may well limit the risk of major failures because adjustments to the reform can be made.

References

Aoki, M. (2001) Toward a Comparative Institutional Analysis. MIT Press, Cambridge, MA, USA.
Aoki, M. (2007) Endogenizing institutions and institutional changes. Journal of Institutional Economics 3 (1): 1-31.
Arthur, W.B. (1988) Self-reinforcing mechanisms in economics. In: P.W. Anderson, K.J. Arrow and D. Pines (eds.). The Economy as an Evolving Complex System. Addison-Wesley, Reading, MA, USA.

[2] NIE has been combined with experimental economics to test the outcomes of different institutional set-ups There is especially a growing literature in the field of using (natural) experiments in development economics (Duflo, 2006). These outcomes may provide information on which institutional patch, in the words of Brousseau and Glachant (2002), may be most worthwhile. An early contribution to this field was Vernon Smith's study of different markets' rules. Experiments can analyse rules and beliefs that structure the behaviour of transactors. With this, there is a clear link to behavioural economics. However, Ménard (2001) warns that too much of experimental economics to date has been devoted to the exploration of the assumptions of rational choice theory. Also, game theory is often used. Game theory essentially analyses strategies and decisions under different circumstances (If person A is expected to do this, I will choose action X, which would then give me the highest payoff).

Brousseau, E. and J. Glachant (2002) The economics of contracts and the renewal of economics. In: E. Brousseau and J. Glachant (eds.). The Economics of Contracts: theories and applications. Cambridge University Press, Cambridge, UK, pp. 3-45.

Coriat, B. and G. Dosi (1998) The institutional embeddedness of economic change: an appraisal of the "evolutionary" and "regulationist" approach. In: K. Nielsen and B. Johnson (eds.). Institutions and Economic Change. Edward Elgar, Cheltenham, UK, pp. 347-76.

Denzau, A.T. and D. North (1994) Shared mental models: ideologies and institutions. Kyklos 47 (1): 3-31.

Duflo, E. (2006) Field experiments in development economics. Report prepared for the World Congress of the Econometric Society. BREAD, CEPR, NBER (January).

Eggertsson, T. (2005) Imperfect Institutions: possibilities and limits to reform. University of Michigan Press, Ann Arbor, MI, USA.

Geels, F.W. (2004) From sectoral systems of innovation to socio-technical systems: insights about dynamics and change from sociology and institutional theory. Research Policy 33 (6-7): 897-920.

Geels, F. and R. Raven (2006) Non-linearity and expectations in niche-development trajectories: ups and downs in Dutch biogas development (1973-2003). Technology Analysis and Strategic Management 18 (3&4): 375-92.

Greif, A. and D.D. Laitin (2004) A theory of endogenous institutional change. The American Political Science Review 98 (4): 633-52.

Hayek, F.A. (1945) The use of knowledge in society. The American Economic Review 35 (4/Sep.): 519-30.

Hendriks, C.M. and J. Grin (2007) Contextualizing reflexive governance: the politics of Dutch transitions to sustainability. Journal of Environmental Policy and Planning 9 (3): 333-50.

Loorbach, D. and J. Rotmans (2006) Managing Transitions for Sustainable Development. Understanding Industrial Transformation: views from different disciplines. Springer, Dordrecht, the Netherlands.

Ménard, C. (2001) Methodological issues in new institutional economics. Journal of Economic Methodology, Taylor and Francis Journals 8 (1): 85-92.

North, D.C. (1990) Institutions, Institutional Change, and Economic Performance. Cambridge University Press, Cambridge, UK.

North, D.C. (1993) Institutional Change: A Framework of Analysis. In: S.-E. Sjostrand (ed.). Institutional Change: Theory and Empirical Findings. M.E. Sharpe, Armonk, NY, USA, pp. 35-46.

North, D.C. (2005) Understanding the Process of Economic Change. Princeton University Press, Princeton, NJ, USA.

North, D.C., J. Wallis and B. Weingast (2009) Violence and Social Orders: a conceptual framework for interpreting recorded human history. Cambridge University Press, Cambridge, UK.

Chapter 3

Social systems and social engineering: Niklas Luhmann

Kristof van Assche, Martijn Duineveld, Gert Verschraegen, Roel During and Raoul Beunen

3.1 Introduction

This chapter introduces a number of concepts from Niklas Luhmann's social systems theory as they relate to innovation, transition and transition management. An understanding of Luhmann's ideas on innovation and steering is essential to grasp a Luhmannian view of system innovation and transition management. Two levels of analysis are developed. The first level centres on the political system, pushing innovation and trying to manage transition, while the second level focuses on organisations and their attempts to innovate. Luhmann is eminently useful in relating the two levels and thus in laying the groundwork for a theory of innovation and transition. An analysis of the development of Dutch discourse on systems innovation, social engineering and transition management since the 1990s serves to illustrate and apply the social systems perspective. Finally, the chapter argues that modernist notions of steering pervading the governance system overestimate the role of governmental actors and underestimate other sources of innovation and systemic innovation that could be labelled 'transition'.

3.2 Social systems: an introduction

Niklas Luhmann (1927-1998) has been described as the greatest social scientist of the 20th century, but many of his concepts have been remarkably unexplored and left without much application. One of the fields in which his insights did gain influence is in organisation and management theory. The following analysis of innovation and transition owes a debt to the work conducted in that particular field of application (e.g. Seidl, 2005; Hernes and Backen, 2002; Teubner, 1996; Fuchs, 2001), allowing for a smoother transition from grand theory to analyses of practice.

Luhmann perceived society as a collection of interacting social systems, and he saw social systems as systems of communication. Each social system creates its own reality through communication, based on specific distinctions and specific modes of reproduction. Systems are operationally closed: everything happens in system conditions, according to the logic of the system. Observations of environments use the logic and semantics of the system itself. And the influence of the environment is only indirect, through interpretation.

Luhmann distinguishes three types of social systems: interactions, organisations and function systems. *Interactions* are conversations, implying the perceived physical presence of interlocutors. They are short-lived systems, fleeting and limited in their processing of environmental complexity (Luhmann, 1995). *Organisations* reproduce themselves through a specific form of communications, namely decisions, implying the awareness and communication of alternatives and continuous reference to previous decisions (Seidl, 2005). *Function systems* are the systems of communication that fulfil a function in society at large.

Luhmann's work looks into the interaction between social systems (organisations or function systems) and psychic systems (individuals) in the production of change and of innovation. Function systems, such as law, economy, politics, religion, science and education, play a role in the reproduction of society as the encompassing social system, each reproducing itself through distinct codes, each maintaining a boundary vis-à-vis the other function systems. The pattern of interdependencies between the function systems reveals a history of mutual adaptation (Van Assche *et al.*, 2010). Politics is considered the system that articulates and enforces collectively binding decisions, but it relies on law for its codification and enforcement (Luhmann, 1990).

Society, for Luhmann, is polycentric in the sense that each function system internally produces an image of society and of the other systems in its environment. Politics is not a site with a superior viewpoint, an elevated position that allows for a comprehensive view and understanding of society (King and Tornhill, 2003; Borch, 2005). Certainly it is not a central position from which the other systems can be steered (distinguishing Luhmann's theory from the high modernism of Scott (1998)). The behaviour of function systems and organisations is necessarily opaque and unpredictable for politics, since politics can never entirely grasp their mode of reproduction. Luhmann, drawing on work in evolutionary biology by Maturana and Varela (1987) calls this their *autopoiesis*, a form of reproduction entirely relying on what is available in the system itself. Autopoiesis creates a matrix of path dependencies within the boundaries of the social system. Direct interference of the environment would halt the autopoiesis, and lead to the dissolution of the system in the environment (Luhmann, 1995, 1989, 1990). In Luhmann's theorising, systems change all the time. These changes and their effects, labelled afterwards as, for example, 'innovation' or 'transition' are declared important, successful, wide-ranging or structural. This labelling implies selections and distinctions made through operations of the system. All of this takes place in a functionally differentiated society, which is itself a product of evolution, leading towards more complex interdependencies.

From this brief introduction, it is clear that Luhmann's systems perspective can be useful in delineating the positive functions of government and governmental organisations in stimulating innovation, and equally so in delimiting those functions. We will now introduce the Dutch discourse on transition and innovation. This discourse assumes considerable governmental steering power. Next, we will analyse the Dutch discourse in the Luhmannian framework and provide a detailed analysis of his views on innovation and transition, simultaneously refining our applied systems perspective. In the concluding section, we assess the Dutch case and reflect on the utility of the theoretical framework.

3.3 System innovation and transition management in the Dutch landscape

In the Netherlands, notions of innovation, systems innovation and transition have been abundant in policy-related documents since the late 1990s. Even the mass media has at times shown a preoccupation with innovation, or a supposed lack thereof. A national innovation forum was convened for a few years, and research promising to enhance innovation was made high priority (Duineveld *et al.*, 2009). Several governmental organisations were retooled to stimulate innovation (Innonet, Habiforum, Ruimtelijk Planbureau and others). The underlying assumptions were usually that the Dutch economy was not sufficiently innovative, not quick enough to adapt to changing environments, that innovation can be measured, that it is something positive, that it can be managed, that rules can be identified to stimulate innovation and manage transition, and, not unimportant, that various branches and levels of government could and should play a role in this urgent process of updating society (Duineveld *et al.*, 2009; Van Assche, 2004, 2006).

System innovation in the Dutch policy context is often used as distinct from innovation. A system innovation could lead to a *transition*, and is supposedly not confined to something small. 'To what then?', we may ask. Different options arise here: to one company, to one sector, to the government, to the business sphere, to academia. This leaves 'innovation' to be defined as 'something new and something good', 'system innovation' to be defined as 'a comprehensive innovation, bridging boundaries', and 'transition' as 'the move of a larger whole by means of system innovation to a higher state of functioning' (cf. Hendriks and Grin, 2007; Avelino and Rotmans, 2009).

A scientific literature developed, sponsored by governmental and quasi-governmental organisations, to circumscribe these concepts more precisely, to define parameters and to generate procedures for their optimisation (e.g. Rotmans, 2000, 2003; Rotmans *et al.*, 2005; Hendriks and Grin, 2007; Kemp *et al.*, 2007; Poppe *et al.*, 2009; Loorbach, 2007). Social and political relevance were considered to be just as important as scientific relevance. Scientific research, it was said, ought to be useful beyond the academic domain, for formulating more effective policies or for strengthening policy implementation. In this context, transition management discourse emerged as an example of a self-perceived successful type of strategic and applied policy and management research.

Within transition management studies, transition is said to consist of different phases, each characterised by its own dynamics and determined by system changes at different scale levels (Loorbach, 2007). A transition is viewed as a process involving several people and organisations, such as ordinary citizens, governments, businesses and social organisations

(Rotmans, 2000). Social actors are considered to be reflexive and as such to shape and influence the dynamics of the system they inhabit (Avelino and Rotmans, 2009).

It is believed that research on long-term societal changes can provide tools that help guide society towards innovation, system innovation for a more sustainable society (Loorbach, 2007). Transition studies carry and constantly reproduce the promise of contributing to solving socio-political and environmental problems. These are represented as complex and persistent dilemmas that have been around for decades for which there are no cut and dried solutions. Such problems are persistent because they are deeply rooted in our social structures and institutions (Rotmans *et al.*, 2005). In order to solve these 'persistent problems' in, for example, agriculture, water management, transportation, education, health care and so on, transition researchers produce recommendations and derive policy measures for managing a transition.

3.3.1 The possibility of steering

Transition management is believed to offer a conceptual framework that enables one to come up with a specific mix of ways to steer things in the right direction. In addition, the transition literature poses that management at the system level is essential, newcomers should create a new regime, a pluralistic approach is desirable and that it ought to be important for the actors involved within transitions to get to know each other's perceptions of reality. More specifically, recommendations are made for setting up a transition arena and developing transition coalitions and a transition agenda. One of the key outcomes of transition research is that 'cooperation' is needed, and that knowledge needs to be more 'interdisciplinary' and 'transdisiplinary' (bridging disciplinary, conceptual and academic boundaries, enabling innovations crossing those boundaries) (Kemp *et al.*, 2007). This framework of assumptions can be easily identified in many research and policy projects.

What is remarkable about the manner in which socio-political changes are represented within transition studies, is the often non-reflexive, latent belief in the possibility of steering transitions or 'social engineering'. Transitions are represented as a set of factors or conditions that, if they all work together, will cause a desired change – as if they are the result of more or less mechanical, instrumental processes. However, according to transition experts it is a misconception to presume that the implementation of the theory will lead to a deterministic collection of directing rules and to a linear production of desired effects (Rotmans *et al.*, 2005).

3.4 Luhmannian responses to the Dutch transition management discourse

We now examine Luhmann's perspective on the transition management discourse, which is sometimes complementary to transition theory and sometimes necessarily deconstructive. Central to this discourse are four notions: (1) innovation and system innovation, (2) the perceived possibilities and limits of steering, (3) the role of the government and (4) the relation between science and transitions. These notions structure the next four subsections.

3.4.1 Luhmann on innovation, system innovation and transitions

What would constitute an innovation from a social systems perspective? What would be a system innovation? For Luhmann, systems must innovate to survive. Social systems reproduce themselves through recursive communication, linking back to previous communications and continuously reinterpreting them. For organisations, each decision reinterprets the history of previous decisions, slowly changing the interpretive frameworks to understand that history, and with it, the image of the organisation that guides further decision making (Luhmann, 2000; Hernes and Bakken, 2002). In autopoietic systems, literally everything will change over time, as structures, elements and procedures gradually transform one another. So, Luhmann's evolutionary theory places a premium on innovation, as something that is radical and necessary for survival (Van Assche, 2006). Yet that continuous change is not always perceived as innovation within the system; and the environment of a system cannot *force* innovation, since direct penetration of the environment in a system would lead to dissolution of the system.

From a systems perspective, it is impossible to distinguish change from innovation. Change takes places continuously, through adaptations to external and internal environments. Innovation could be seen as important change, or as radical change, rapid change or successful change, but these labels have little theoretical relevance. A distinction that system theory can make is that between *conscious and unconscious change.* Social systems are self-aware and capable of self-reflection, that is, with communication on communication and communication on its position vis-à-vis various environments (Luhmann, 1989, 2004). But they never have a full understanding of their own autopoiesis, since they necessarily observe from within, that is, with all the blind spots of first-order observation (Seidl, 2005; During *et al.*, 2009).

Social systems, therefore, will be unable to reflect on all changes taking place in the system, and many of those changes are unconscious adaptations. Conscious changes are changes that enter the self-reflection of the system, as conscious responses to observations of the environment. If a conscious change is deemed successful, it can be labelled as an innovation *in hindsight* (Seidl and Becker, 2005). A response to a changing environment might be inspired by fear, and marked by erratic analyses of skills, demands and resources, but if it works, it might become an innovation. As said, the chain of reinterpretations of success and failure, of innovation and non-adaptation never stops (Morgan, 1986; Jansson, 1989; Brunsson, 2002; Van Assche, 2004).

Rhetoric is the art of saying things well and the art of being persuasive. The rhetoric of innovation, though, is not restricted to results of conscious responses. If an organisation changes and is largely unaware of those changes, and the result is perceived as very positive (e.g. an unexpected success in elections, a growing market share, a marvellously implemented plan) then the rhetoric of innovation can still claim that success *a posteriori*, as the result of conscious decision making, of a drive for innovation (Beunen *et al.*, 2009; Allino-Pisani, 2008; Ledeneva, 2005; Collins, 2006). Lack of innovation can be observed only by a collapse of autopoiesis, which in the case of a company could take the form of a bankruptcy.

The fact that innovation and, by extension, system innovation is also a matter of semantics derives from the *self-referential* character of a system's reproduction. Vos (in Seidl and Becker, 2005) distinguishes, with Luhmann, three forms of self-reference involved in autopoiesis: (1) basal or operational self-reference, which is the recursive relation between communications, (2) *reflexivity*, which is conscious reference to the communication process, and (3) *reflection*, which is communication about the system/environment distinction. Through reflection, organisations make sense of their environment and of expectations of that environment, for example, of customers and competitors. Taking these three types of self-reference together, the implications of operational closure for perceptions of and adaptations to the environment become clearer. Every hetero-reference and every reference to an environment is necessarily grounded in self-reference (Van Assche, 2010). Each image of the environment, every formulated response to change, is the product of self-reference. In a reformulation of the radically constructivist character of Luhmann's theory, we can say that organisations are *built on nothing;* there is no ultimate ground for the decisions they take. The reasons for their decisions, and for innovative changes in their mode of reproduction, including images of self, environment, ascriptions of cause and effect, success and failure, are all products of the system itself *and could have been constructed otherwise* (Luhmann, 2000; Hernes and Bakken, 2002; Seidl, 2005).

This is what Luhmann calls the paradoxical nature of decision, and hence, of organisations. If organisations were fully aware of the nothingness upholding them, autopoiesis would stop, so a series of strategies evolved to hide the paradox from themselves. Luhmann speaks of *de- paradoxification* (Teubner, 1996; Van Assche, 2006). The paradox cannot be eliminated or solved, but it can be shifted out of sight (Schiltz, 2007). Causalities are constructed: 'we have to do this'; and leadership qualities are mystified: 'she (or he) will guide us through this'. Organisations construct environments and events in such a way that they seem to cause, or prompt, a certain decision (Hernes and Bakken, 2002; Weick, 1995; Czarniawska, 1997). In this rhetoric of de-paradoxification, certain features of the external or internal environment are constructed as inviting, enabling or requiring innovation (During *et al.*, 2009). Strategic management of organisations, including innovation management, cannot but be self-referential.

Innovation can take place at the level of communications, procedures and structures, but changes in one feature of the system are enabled by changes in the others, and will trigger new adaptations there. Changes in self-image, images of the environment, in decision procedures and in hierarchy all interrelate, whether one is aware of this or not (King and Tornhill, 2003). In that sense, every innovation is a system innovation; it is just that some changes are labelled systemic, and others not. One can refer to the American automobile industry, with each of its successive waves of reform presented by successive managers as more systemic and more innovative than the previous ones.

What could be a *transition* in Luhmannian terms? Situations can be labelled transitional when it is opportune to do so, and when there is some awareness of systemic changes. Awareness of a transition can be linked to a rhetoric of transition, or not. There can be an awareness of a goal, or not. There can be a clear plan, a clear route towards that goal, or not (Van Assche *et al.*, 2010). Post-communist countries, long called 'transitional', were all supposedly similar and supposed to end up similar. But practice showed that the specific autopoiesis of each state organisation, their specific internal and external environments and their informal institutions, produced very different effects, and those same factors also produced different images of routes, plans and final situations (Elster *et al.*, 1998). Post-communist 'transition' also demonstrated the false assumptions under which many western consultants were operating. These were products of their own organisational cultures and business models, which produced assumptions about ready-made models for capitalist democracies and simple interventions to build them (Allina-Pisano, 2008; Verdery, 2003; Ledeneva, 2005). The same example reveals the many ways that the rhetoric of transition can be internally appropriated. Groups stress different aspects of the final situation, and the same applies to the starting point and the path. Some participants work towards a final goal with conviction, while others just adopt the rhetoric for various purposes, criminal or otherwise (Collins, 2006).

3.4.2 Luhmann on the limits of steering

The success of Dutch transition studies is partly explained by the promise they hold, which perfectly fits the quest for innovation and system innovation. In practice, however, the usefulness of these recommendations is often limited. There are simply too many factors that influence the implementation of policies and therefore these policies will rarely work out as intended and expected (Pressman and Wildavsky, 1979). Luhmann is critical too about projects of social engineering. He criticised in various works, among them *Political Theory in the Welfare State* (1990), attempts of the state to work towards the ideal society. He doubted the assumption that the political system, in conjunction with the bureaucracy, would be able to gain an overview of society and of all social systems, along with their problems and their smooth and not-so-smooth interactions. One can link this lack of observational capacity in the political system to a similar lack within every organisation. The more radical the intervention attempted and the more unpredictability introduced, the higher the risk for both intervening and subjected system (Luhmann, 1989), and the greater the pressure for the intervening, the 'managing' system, to keep managing, to expand its operations and to overburden itself with regulatory tasks (Luhmann, 1990, 2000; Van Assche and Verschraegen, 2008; Van Assche and Leinfelder, 2008).

Steering or engineering transitions is all the more difficult, because of the embedding of organisations in various function systems (Hernes and Bakken, 2002). Whereas politics cannot steer the other function systems without introducing more unpredictability and less differentiation, the steering problem is compounded by the complex relationships between organisations and function systems (Seidl and Becker, 2005). Organisations, unlike function systems, can receive the ascription 'actor', that is, they can be addressed in communication (Luhmann, 2002). Thus, communication between governmental organisations and other actors is possible, whereas communication between function systems is not (Luhmann, 1995). This would seem to open the door for direct steering of organisations by their governmental peers. However, the requirement of operational closure still stands, and the unpredictability of organisational response is increased by the unpredictable influences of law, economy and education. A consultancy firm, for example, can be pushed towards innovation by governmental policies, but it cannot afford to ignore the financial bottom line, and it also participates in the systems of science and education, opening the organisation to the autopoietic requirements of those systems (Simon, 2002; Jansson, 1989; Van Assche, 2004). Indeed, tensions within organisations routinely emerge from the conflicting requirements of different function systems, say science and economy for the consultancy (Seidl, 2005). But a push towards innovation, that is, a push towards systemic change in the organisation, will likely change its functional embeddings, aggravate those tensions and shift its responses to future directives (Duineveld *et al.*, 2007).

3.4.3 Luhmann on government

For Luhmann, welfare states' overstepping the boundaries of politics and overestimating their steering power create their own disappointments (Luhmann, 1989, 2000; Willke, 1994; Van Assche, 2006). Failed policy results in calls for new policies and new attempts to intervene in the other systems (Beunen *et al.*, 2009). This increases the tasks of the observing system, increases the complexity of the observations and procedures required, and raises the difficulties of managing its own internal complexity (Hernes and Bakken, 2002). Typically, this implies a proliferation of bureaucracy, simultaneously slowing down the processing of environmental information, and consequently, widening the gap between observation and policy response, resulting in a more outspoken blindness to the other systems (Luhmann, 1989, 1990).

In welfare states, and certainly in communist regimes, the degree of functional differentiation decreases because of recurring interventions by the political system, because of semantics in politics placing itself at the centre of society (Sievers, 2002; Elster *et al.*, 1998; Van Assche *et al.*, 2010). When politics tries to take over law, economics and education, those systems will gradually lose their operational closure and lose the capacity to autopoietically reproduce. After a while, this leads to a breakdown of the systemic logic (King and Tornhill, 2003). This, in turn, means a loss of observational capacity for society as a whole, which is precisely the multitude of *different* observations in various function systems that allows society to adapt to ever changing environments (Van Assche, 2006, 2010). This is what Luhmann terms *de-differentiation* (1989, 1995, 2000).

3.4.4 Luhmann on the relation between transition management and science

Luhmann (2002: 218) claims that 'we have developed a society that has no choice but to run risks'. Differentiation makes society dependent on a collection of systems that cannot adequately predict each other's behaviour: a risky operation. One response, according to Luhmann, is a growing dependence in the autopoiesis of society on *second-order observation*, that is, the observation of observations. One of the more pleasant consequences of relying on second-order observation, which includes but is not restricted to scientific analysis, is that risk can more accurately be made visible, enabling more precise risk management. Since innovation, and by extension transition, implies increased risk and since much of the behaviour of an organisation is as yet untested, risk management becomes a more central concern for the organisations involved (cf. Douglas and Wildavsky, 1982).

This opens the door to a stronger reliance on science, a strategy that brings its own risks. Scientific observation, grounded in the distinction true/false, and conditioning truth in the application of scientific method and theory, can analyse risky and innovative decision making in other systems, in second-order observation, but it brings its own blind spots. And, being a social system itself, it cannot fully grasp the complexity of economic, political and legal decision making. Political pressure on science to find *the* logic of innovation, born from a desire for economic development, a logic that cannot be discovered following the logic of science itself, can lead to recipes for innovation and transition that misleadingly receive the stamp of science (Latour, 1987, 2004; Van Assche, 2004, 2006; Duineveld *et al.*, 2007, 2009). If factored into the risk management of organisations, these become recipes for disaster (Luhmann, 1989, 2002; Douglas and Wildavsky, 1982).

Governmental incentives for innovation and transition can further be recuperated by the other actors for other reasons. Subsidies can lead to re-labelling existing knowledge, products and procedures as new and create unfair competition in business, science and education (Latour, 2004; Duineveld *et al.*, 2007; Van Assche, 2006). Better than trying to grasp the reproduction of all other systems and trying to invent context-independent formulas for innovation is what Willke (1994) calls 'context guidance'. Creating improved conditions for innovation and systemic innovation, in all likelihood has to entail a redistribution of risk. In other words, this enables innovation by communal absorption of risks taken by innovative actors. Yet this strategy has limits too, as sufficiently illustrated by the recent crisis of innovative banking.

3.5 Conclusions: limits and possibilities of steering transitions

The transition management discourse emerged in the Dutch context, rooted in a belief in the possibilities of social engineering and a firm entanglement of politics, science and education. In an ideal differentiated society, companies should think and act as companies, scientific organisations as scientific organisations, and administrations as administrations, assisting politics in taking and implementing collectively binding decisions. Assuming that the government can identify a non-existing logic of innovation and transition, taking the next step to try to impose that on a society deemed transparent, inevitably endangers the autonomy and operational closure of other social systems. This, in turn, jeopardises the gains of differentiation: it irrevocably reduces the complexity of models of the environment produced in society and therefore the quality of adaptations to that environment (Luhmann, 1989, 1990). In other words, it makes innovation and transition less likely.

Luhmann would argue that the practice and the semantics of innovation are specific to each system, as are the available pathways. Path dependencies are manifold, and they are specific to different organisations and different function systems (Van Assche *et al.*, 2010). A scientific innovation is not an economic innovation, which is not a legal innovation. And at any given point in time, the possibilities for those function systems to evolve are constrained by their history of autopoiesis. The same applies to organisations as social systems. The difference between function systems and organisations precludes the possibility of any function system fully understanding the process of innovation in any organisation. Organisations, like companies, can utilise a rhetoric of innovation to impress government, to sell, to provoke pity, to evade taxes and to receive subsidies (Brunsson, 2002; as early as Burns and Stalker, 1961).

Governments have very limited means of assessing those claims, and close association with firms makes them more vulnerable. Innovation in general, as in the successful restructuring of a system to deal with environmental change, or to make use of unseen opportunities, cannot be uniformly described and recognised, it cannot be predicted, it cannot be forced. This is very clear in scientific discovery and in economic change, where success is never predictable (Latour, 1987; Weick, 1995; Czarniawska, 1997). Neither can the sufficient knowledge base to enable success be circumscribed. Nevertheless, recurring modernist ideologies tend to install expectations that innovation is one phenomenon, that it can be engineered, and in social democracies like the Netherlands, that government has the duty at least to contribute to the engineering efforts (Luhmann, 1990; Scott, 1998; Duineveld *et al.*, 2007, 2009; Van Assche, 2004).

All of the problems signalled by politics and administration trying to manage society are intensified when government, supported by scientific misconceptions, tries to standardise innovation and enforce the fictitious standards found. Unfortunately, concepts like system innovation and transition management figure prominently in those attempts. Surely, there are innovations in different social systems that are interrelated, that might be triggered by each other or by somehow similar observations, and one could label some of those related changes system innovations, or transitions towards system innovation. This analysis, however, does not allow for the conclusion that system innovations represent a single phenomenon, obeying one set of laws and open to engineering by a benevolent government.

What can governments do to enable transition? *First of all*, we would argue for examining whether there really is a systemic lack of innovation. Simultaneously, where are the verdicts of a lack of innovation coming from? A critical examination of innovation in various systems conducted through open conversations with organisational actors with differing functional embeddings can greatly enhance the quality of the analysis. Often, innovation is reduced to technical innovation, and successful innovation is essentially measured by economic criteria. *Secondly*, we would argue to safeguard differentiation in society. Differentiation enables refined adaptation, hence innovation. Limit interference in the autopoiesis of the other function systems. *Thirdly*, we propose developing sustained reflection, as second-order observation, on path dependencies and interdependencies in and between function systems and organisations. Such reflection cannot be restricted to science, and scientific reflection cannot be affected by a political desire for simple answers. *Finally*, we suggest keeping the rules of economic competition and scientific achievement as clear as possible, and trying to avoid distortion of incentives within specific social systems by innovation policies trying to orchestrate change processes among social systems.

Luhmann helps us in the analysis of management, of innovation and of the relation between various function systems, specifically between a steering government and the other systems. He also helps us in understanding the delicate balance between self-reproduction and external relationships that allows social systems, and society as the encompassing social system, to evolve and to innovate. No centre is needed for that, no perfect overview of society, no strong steering capacity for politics. All that is required is a set of interdependent yet operationally closed social systems. Society did not start out like this. Our functionally differentiated society is a remarkable evolutionary achievement, and functional differentiation, in a Luhmannian perspective, is surely the most important innovation of modern society. Yet it is such a slow achievement that nobody noticed it, and it is so systemic that no actor has a claim to the fame of it.

References

Allina-Pisano, J. (2008) Post Soviet Potemkin Villages: politics and property rights in the black earth. Cambridge University Press, Cambridge, UK.

Avelino, F. and J. Rotmans (2009) Power in transition: an interdisciplinary framework to study power in relation to structural change. European Journal of Social Theory 12: 543-69.

Beunen, R., W. Van der Knaap and G. Biesbroek (2009) Implementation and integration of EU environmental directives: experiences from the Netherlands. Environmental Policy and Governance 19 (1): 56-73.

Borch, C. (2005) Systemic power: Luhmann, Foucault, and analytics of power. Acta Sociologica 48: 155-67.

Brunsson, N. (2002) The Organization of Hypocrisy: talk, decisions and actions in organizations. Copenhagen Business School Press, Oslo, Norway.

Burns, T. and G. Stalker (1961) The Management of Innovation. Tavistock, London, UK.

Collins, K. (2006) Clan Politics and Regime Transition in Central Asia. Cambridge University Press, Cambridge, UK.

Czarniawska, B. (1997) Narrating the Organization: dramas of institutional identity. University of Chicago Press, Chicago, IL, USA.

Douglas, M. and A. Wildavsky (1982) Risk and Culture: an essay on the selection of technical and environmental dangers. University of California Press, Berkeley, CA, USA.

Duineveld, M., R. Beunen, K. Van Assche, R. During and R. Van Ark (2007) The Difference Between Knowing the Path and Walking the Path: een essay over het terugkerend maakbaarheidsdenken in beleidsonderzoek. Wageningen University, Wageningen, the Netherlands.

Duineveld, M., R. Beunen, K. Van Assche, R. During and R. Van Ark (2009) The relationship between description and prescription in transition research. In: K.J. Poppe, C. Termeer and M. Slingerland (eds.). Transitions Towards Sustainable Agriculture and Food Chains in Peri-Urban Areas. Wageningen Academic Publishers, Wageningen, the Netherlands, pp. 309-325.

During, R., K. Van Assche and A. Van der Zande (2009) Culture, innovation and governance in Europe: systems theories and the analysis of INTERREG programs. In: K.J. Poppe, C. Termeer and M. Slingerland (eds.). Transitions Towards Sustainable Agriculture and Food Chains in Peri-Urban Areas. Wageningen Academic Publishers, Wageningen, the Netherlands, pp. 163-189.

Elster, J., C. Offe and U. Preuss (1998) Institutional Design in Post-Communist Societies: rebuilding the ship at sea. Cambridge University Press, Cambridge, UK.

Fuchs, S. (2001) Against Essentialism: a theory of culture and society. Harvard University Press, Cambridge, MA, USA.

Hendriks, C.M. and J. Grin (2007) Contextualizing reflexive governance: the politics of Dutch transitions to sustainability. Journal of Environmental Policy and Planning 9: 333-50.

Hernes, T. and H. Bakken (2002) Autopoietic Organization Theory. Copenhagen Business School Press, Oslo, Norway.

Jansson, D. (1989) The pragmatic uses of what is taken for granted: project leaders applications of investment calculus. International Studies of Management and Organization 19 (3): 49-63.

Kemp, R., D. Loorbach and J. Rotmans (2007) Transition management as a model for managing processes of co-evolution towards sustainable development. International Journal of Sustainable Development and World Ecology 14: 78-91.

King, M. and C. Tornhill (2003) Niklas Luhmann's Theory of Politics and Law, Houndmills: Palgrave Macmillan.

King, M. and C. Tornhill (eds.) (2006) Luhmann on Law and Politics: critical appraisals and applications. Hart publishing, Oxford, UK.

Latour, B. (1987) Science in Action: how to follow scientists and engineers through society. Harvard University Press, Cambridge, MA, USA.

Latour, B. (2004) Politics of Nature: how to bring the sciences into democracy. Harvard University Press, Cambridge, MA, USA.

Ledeneva, A. (2005) How Russia Really Works. Cornell University Press, Ithica, NY, USA.

Ledeneva, A. (2007) Transition Management: new mode of governance for sustainable development. Erasmus University, Rotterdam, the Netherlands.

Loorbach D. (2007) Transition Management. New mode of governance for sustainable development. Erasmus University, Rotterdam, the Netherlands.

Luhmann, N. (1989) Ecological Communication. University of Chicago Press, Chicago, IL, USA.

Luhmann, N. (1995) Social Systems. Stanford University Press, Stanford, CA, USA.

Luhmann, N. (1990) Political Theory in the Welfare State. Mouton de Gruyter, Berlin, Germany.

Luhmann, N. (2002) Risk: a sociological theory. Aldine Transaction, New Brunswick, NJ, USA.

Luhmann, N. (2004) Law as a Social System. Oxford University Press, Oxford, UK.

Luhmann, N. (2000) Organisation und Entscheidung. Westdeutscher Verlag, Opladen, Germany.

Maturana, H.R. and F.J. Varela (1987) The Tree of Knowledge: the biological roots of human understanding. Shambhala Publications, Boston, MA, USA.

Morgan, G. (1986) Images of Organization. Sage, Beverly Hills, CA, USA.

Poppe, K.J., C. Termeer and M. Slingerland (eds.) (2009) Transitions Towards Sustainable Agriculture and Food Chains in Peri-Urban Areas. Wageningen Academic Publishers, Wageningen, the Netherlands.

Pressman, J.L. and A.B. Wildavsky (1979) Implementation: how great expectations in Washington are dashed in Oakland. University of California Press, Berkeley, CA, USA.

Rotmans, J. (2000) Transities en Transitiemanagement: de casus van een emmisiearme energievoorziening. International Centre for Integrative Studies, Maastricht, the Netherlands.

Rotmans, J. (2003) Transitiemanagement: sleutel voor een duurzame samenleving. Van Gorcum, Assen, the Netherlands.

Rotmans, J., D. Loorbach, R. Van der Brugge (2005) Transitiemanagement en duurzame ontwikkeling: co-evolutionaire sturing in het licht van complexiteit. Beleidswetenschap 19: 3-23.

Schiltz, M. (2007) Space is the place: the laws of form and social systems. Thesis Eleven 88 (1): 8-30.

Scott, J. (1998) Seeing Like a State. Yale University Press, New Haven, CT, USA.

Seidl, D. (2005) Organizational Identity and Self-transformation: an autopoietic perspective. Ashgate, Aldershot, UK.

Seidl, D. and K. Becker (eds.) (2005) Niklas Luhmann and Organization Studies. Copenhagen Business School, Copenhagen, Denmark.

Sievers, E. (2002) Uzbekistans mahalla: from Soviet to absolutist residential community associations. Journal of International and Comparative Law at Chicago Kent 2 (2): 91-150.

Simon, F. (2002) The deconstruction and reconstruction of authority and the role of management and consulting. Soziale Systeme 8: 283-93.

Teubner, G. (1996) Double bind: hybrid arrangements as de-paradoxifiers. Journal of Institutional and Theoretical Economics 152: 59-64.

Van Assche, K. (2004) Signs in Time: an interpretive account of urban planning and design, the people and their histories. Wageningen University, Wageningen, the Netherlands.

Van Assche, K. (2006) Over Goede Bedoelingen en Hun Schadelijke Bijwerkingen. Innovatienetwerk Groene Ruimte, Utrecht, the Netherlands.

Van Assche, K. (2010) The good, the bad, and the self- referential: heritage planning and the quest for productive difference. In: T. Bloemers and A. Van der Valk (eds.). The cultural landscape and heritage paradox: Protection and development of the Dutch archaeological-historical landscape and its European dimension. Amsterdam University Press, Amsterdam, the Netherlands.

Van Assche, K. and H. Leinfelder (2008) Nut en noodzaak van een kritische planologie: suggesties vanuit Nederland en Amerika op basis van Niklas Luhmann's systeemtheorie. Ruimte en Planning 28 (2): 28-38.

Van Assche, K. and G. Verschraegen (2008) The limits of planning: Niklas Luhmann's social systems theory and the analysis of planning and planning ambitions. Planning Theory 7 (3): 263-83.

Van Assche, K., J. Salukvadze and G. Verschraegen (2010) Changing frames: expert and citizen participation in Georgian planning. Planning Practice and Research 25 (3): 377-395.

Verdery, K. (2003) The Vanishing Hectare: property and value in postsocialist Transsylvania. Cornell University Press, Ithica, NY, USA.

Weick, K. (1995) Sense-Making in Organizations. Sage, Thousand Oaks, CA, USA.

Willke, H. (1994) Systemtheorie II: interventionstheorie. UTB, Stuttgart, Germany.

Chapter 4

The role of social networks: Mark Granovetter

Machiel Reinders

4.1 Introduction

Structural transformations or transitions are necessary to resolve social problems. A 'transition' can be defined as a long-term process of change during which a society or a subsystem of society is fundamentally altered (Loorbach and Rotmans, 2006). Transitions require systems of innovations. Typically, this implies the co-evolution of different innovations within technological niches and developments on an exogenous (societal) level that put pressure on socio-technical systems, forcing them to change (Geels, 2004). However, socio-technical systems do not function autonomously, but are the outcome of the activities of human actors who are embedded in social groups. Though social groups have their own characteristics, the fact that groups are often mutually interdependent and are characterised by interpersonal interaction is an important notion in transition theory (e.g. Geels, 2004). As this chapter will show, the analysis of these interaction processes in interpersonal networks is helpful in relating micro-level (niche-level) developments by individual actors to higher level patterns (like socio-technical regimes and landscapes).

Mark Granovetter (1973) developed a theory with which the strength of interpersonal ties in small-scale interactions can be used to explain macro phenomena, such as diffusion of innovations, social mobility, the organisation of communities and social cohesion in general. As such, the strength-of-ties perspective can be helpful in explaining facets of transitions or changes in socio-technical regimes.

This chapter starts by providing some background information about the strength-of-ties theory. It then tries to relate both theoretical perspectives, and more specifically, to explore what strength-of-ties theory can add to the literature on transition management. Tensions between the strength-of-ties perspective and the assumptions of transition theory are then examined and some prospects for action are suggested.

4.2 Strong and weak ties

The strength of interpersonal or dyadic ties can be defined as the combination of 'the amount of time spent in interaction, the emotional intensity, the intimacy, and the reciprocal services which characterise the tie' (Granovetter, 1973: 1361). Stated differently, tie strength characterises the closeness and interaction frequency of a dyadic relationship. Individuals are sometimes influenced by others with whom they have tenuous or even random relationships (e.g. acquaintances). These influences are labelled 'weak ties'. People also engage in more stable, frequent and intimate 'strong tie' interactions (e.g. with close friends). According to the 'strength of weak ties' theory, weak social ties are more likely to link members of different groups than are strong ties (Granovetter, 1973; Brown and Reingen, 1987). If these weak ties did not exist, a system would consist of disjointed subgroups, inhibiting the widespread diffusion of information (Brown and Reingen, 1987).

Weak ties thus fulfil a bridging function between different social networks. As such, they are assumed to be responsible for the dissemination of information. Given this assumption, tie strength can be used to explain various phenomena. An example is the diffusion of innovations and ideas. Ideas, innovations or whatever is to be diffused can reach a larger number of people and travel a greater social distance when passed through weak ties rather than strong ones. When the diffusion of a new idea only goes through strong ties, it is much more likely to be limited to a few cliques than when it moves through weak ties. So, individuals with many weak ties are best placed to diffuse innovations or ideas, since some of those ties will form local bridges between different networks. Another application of strength-of-ties theory is to social mobility. For example, when one changes jobs, he or she not only moves from one network of ties to another, but also establishes a link between these. Especially within professional and technical clusters that are well defined and limited in size, such social mobility connects clusters with bridging weak ties. Taken together, weak ties can be said to play a role in effecting social cohesion.

Next to the original idea of the strength of weak ties creating social cohesion between separate networks or clusters, more recent literature points out other advantages of weak ties. In addition, there is a literature that counterargues why strong ties still fulfil an important role in social networks. We elaborate on these studies shortly. First, Granovetter (1973) theorised that weak ties (those typified as distant and with infrequent interaction) are more likely to be sources of novel information, because strong ties tend to be connected to others who are close to the knowledge-seeker and thus possess information that the seeker already knows. To put it another way, overlapping or redundant knowledge is the product of social actors sharing equivalent structural positions in which they are exposed to similar types of information (Burt, 1987, 1992). This is often the case for actors sharing strong ties, while weak ties are characterised by a low degree of knowledge redundancy.

Whereas weak ties appear very well suited for the transfer of new knowledge, they seem to be less suited for transferring tacit or complex knowledge. Tacit knowledge is difficult to transmit to another person in writing or by verbalising it, and it is not easily shared. Tacit knowledge often consists of habits and cultural traits that we do not recognise in ourselves and are unaware that we possess. We therefore often overlook how it could be valuable to others. Effective transfer of tacit knowledge generally requires extensive personal contact and trust. Transferring this type of knowledge is less difficult through strong ties, because the actors understand each other and allow for a two-way interaction that is absent in weak ties (Hansen, 1999).

Uzzi (1997) showed that strong tie relationships are effective because they tend to be trusting and helpful in solving problems. They constitute a base of trust that can reduce resistance to change and provide comfort in the face of uncertainty (Krackhardt, 1992). So, people in insecure positions are more likely to resort to strong ties for protection and to reduce uncertainty (Granovetter, 1982). Strong ties have also been claimed to be more accessible and willing to be helpful (Krackhardt, 1992). Table 4.1 provides an overview of the advantages of weak and strong ties based on a short review of tie strength studies in management literature. As we will see, these different perspectives on tie strength point to interesting ways to look at transitions.

4.3 Applying strength-of-ties theory to the transition literature

There are several linkages between the strength-of-ties perspective and the transition literature. In particular, this section focuses on three aspects of transition theory to which the strength-of ties perspective can be applied: (1) dynamic interactions between actors, rule

Table 4.1. Characteristics of strong versus weak ties.

Strong ties	Weak ties
Provide familiar, redundant information[3,4,6]	Provide new, nonredundant information[3,4,6]
Necessary for transfer of complex knowledge[4]	Necessary for transfer of simple knowledge[4]
Necessary for transfer of tacit knowledge[4]	Necessary for transfer of explicit knowledge[4]
High levels of trust[5,7]	Lower levels of trust[5,7]
High accessibility[5]	Lower accessibility[5]
Willing to be helpful[2,5,7]	Instrumental[5,7]
Conducive to the flow of influence[1]	Conducive to the flow of information[1]
Share sensitive information[6]	Provide access to a greater amount and diversity of information[6]
High relational embeddedness[6]	Low relational embeddedness[6]

[1] Brown and Reingen (1987).
[2] Cross and Sproull (2004).
[3] Granovetter (1973).
[4] Hansen (1999).
[5] Levin and Cross (2004).
[6] Rindfleisch and Moorman (2001).
[7] Uzzi (1997).

regimes and systems; (2) the relations between different levels where change takes place; and (3) the management of transition processes.

4.3.1 Dynamic interactions between actors, rule regimes and systems

We distinguish dynamic interactions between actors and rules and between actors and socio-technical systems. First, actors and organisations are embedded in interdependent networks with shared sets of rules and mutual dependencies. Actors interact within the constraints and opportunities of existing rules, but at the same time they act upon and restructure these rule regimes. Through social actions, new shared rules are created (e.g. in the form of routines and norms). However, social learning is required to create new rule regimes and to effectuate the new shared rules. Social learning takes place through imitation or the exchange of experiences. Weak ties play a crucial role in social learning, as they bridge different social networks.

Second, socio-technical systems do not function autonomously, but are the outcome of the activities of human actors (Geels, 2004). On one hand, these systems are maintained and changed by the activities of actors, whereas on the other hand, socio-technical systems also form a context for actions. Socio-technical systems change because actors react to each other's strategic moves. Again, weak ties play an important role here. They provide interactions between different groups (e.g. public and private parties), making it possible to react to each other's strategies. As a consequence, practices and products as established in socio-technical systems are changed.

In sum, the strength-of-ties perspective might help to explain why interactions take place between actors, on the one hand, and between systems and rules on the other hand. More specifically, weak ties allow learning and imitation between actors in different networks but operating within the same socio-technical regime. They are a key factor in the process of the continual reshaping of existing rules and systems.

4.3.2 Relations between different levels

The multi-level concept of transitions implies that transitions can be realised when developments interfere with and reinforce each other in the same direction at three levels (the technological niche, the socio-technical regime and the landscape) (see e.g. Geels, 2004; Geels and Schot, 2007; Rotmans *et al.*, 2005). Niches are important for the emergence of radical innovations that might eventually help transitions to take place. Niches tend to be clusters of relatively strong ties. They are often well defined and limited in size, giving them a 'sense of community'. However, according to Geels and Raven (2006) experimental projects in these niches provide space for interactions with actors from other niches and facilitate the building of social networks. Through such experimental projects, information and ideas flow more easily through professional and technical niches activated at meetings and conventions.

Moreover, because niches are locations that are protected from the mainstream market and where it is possible to deviate from the rules in the existing regime, learning and imitation between different experimental local projects also takes place. So, these experiments are helpful in bridging social networks and establishing new weak ties. These weak ties are instrumental in helping different local projects to gradually add up to a technological trajectory at the global (or landscape) level.

4.3.3 Management of transitions

Finally, the strength-of-ties perspective can be applied to transition management. Transition management aims to better organise and coordinate transition processes (Loorbach and Rotmans, 2006). Transition arenas play an important role here. Transition arenas are networks of innovators and visionaries that develop long-term visions and images that provide the basis for the development of agendas and experiments. Various parties are brought together in these transitions arenas. As such, they form a network of weak ties through which new knowledge is brought together and exchanged. Weak ties thus again facilitate the exchange of knowledge between different social networks or actors.

In each of these three specific aspects of transition theory, the strength-of-ties perspective can be applied. Overall, we can conclude that the success of transitions is conditional on the linkage of otherwise disconnected social networks. Strength-of-ties theory helps to explain the mechanism by which these networks are linked. However, as we will see next, there are also some tensions between the theoretical perspectives.

4.4 Tensions between both theoretical perspectives

Here, we elaborate on two aspects of strength-of-ties theory that interfere with the assumptions of transition theory: (1) the ways in which knowledge travels through networks and (2) trust between actors.

First, as mentioned earlier, weak ties have the advantage of connecting disjointed networks, but they have difficulty transmitting complex knowledge (see e.g. Hansen, 1999). Though a social system comprised of weak ties can profit the most from the dissemination of knowledge, severe transfer problems occur when the knowledge is complex. Strong ties are embedded in the same network and share the same ideas, concepts, norms and values. For example, organisations have shared norms and beliefs and unwritten practices and traditions which are difficult to formulate and transfer (tacit knowledge). Weak ties are helpful in bridging social networks and allowing the interactions between different niches that are necessary to effectuate transitions. However, the transfer of complex or tacit knowledge still forms a problem. More specifically, radical innovations encompass major improvements over existing products or processes; they therefore benefit from complex (i.e. tacit) knowledge transfer (Wuyts *et al.*, 2004: 90; Iansiti and West, 1997; Zucker *et al.*, 2002). Only frequent and repeated interaction facilitates the transfer of tacit knowledge, generates a deeper understanding of new technologies and induces shared mental models (Dewar and Dutton, 1986; Madhaven and Grover, 1998; Wuyts *et al.*, 2004). Thus, while strength-of-ties theory emphasises the difficulty of exchanging complex knowledge through weak ties, this is often a crucial aspect in transition processes.

The second aspect is the role of trust. Trust is essential in the exchange of knowledge necessary for the success of innovations (Bouty, 2000). According to Levin and Cross (2004), trusting an actor as knowledge source to be benevolent and competent should increase the chance of the other party learning from the interaction. As such, trust is a necessary condition in effectuating transitions. Theory states that weak ties are characterised by a lower level of trust, and relationships between weak ties are more instrumental and based on the arm's length principle, whereas relationships through strong ties are based on commitment, trust and the willingness to invest in and help each other (see e.g. Uzzi, 1997 and also Table 4.1). Based on this, the interaction and cooperation between social networks that is important to make transitions work, often have a fragile basis characterised by opportunism and actors' own interests.

These critical notes are relatively underexposed in transition theory. Looking at transitions from a social network perspective helps us to see transition processes from a human point of view, including the shortcomings when people interact or communicate with one another.

4.5 Perspectives for action

This chapter has looked at transition theory from a strength-of-ties perspective. Social network theory can complement transition theory in explaining how the interactions between different social networks allow learning and imitation and might help us to stimulate the transfer of new knowledge. The discussion in the previous sections suggests some useful considerations and perspectives for action.

We argued that weak ties are a key factor in establishing social cohesion between separate networks or clusters and that weak ties are more likely to be sources of novel information. Yet, weak ties are less suited for the transfer of tacit or complex knowledge. Moreover, Uzzi (1997) showed that strong tie relationships are effective because they tend to be trusting and helpful in solving problems. Solutions which combine the benefits of strong and weak ties could provide guiding principles for effectuating transitions. For example, Levin and Cross (2004) show that individuals and organisations benefit from developing trusted weak ties. They argue, 'Our finding on the benefits of perceived trustworthiness plus weak ties seems particularly promising for practitioners in light of the fact that prior research has suggested that weak ties may also be less costly to maintain' (Levin and Cross, 2004: 1487). With regard to establishing trust between weakly linked actors, emphasising common interests is very important to keep all parties involved. A solution to the problem of complex knowledge transfer through weak ties is to intensify meetings among the weakly tied parties at which knowledge can be exchanged. However, this could lengthen the transition process.

Cross and Sproull (2004) showed that strong and weak ties can be complementary. Since weak ties are instrumental in providing new information, weak ties are positively related to the receipt of solutions. Since strong ties are useful in the transfer of tacit or complex knowledge, strong ties are positively related to problem reformulation and validation. Stated differently, strong ties may shape how people think about problems, and thus *the way* people seek additional information, while weak ties could provide this additional information and help generate new ideas and solutions. Therefore, both types of ties are needed to facilitate societal change.

Finally, Rindfleisch and Moorman (2001) showed that in an inter-organisational context relational embeddedness does not necessarily lead to knowledge overlap, as is traditionally the case for strong ties. For example, members within a supply chain have a relatively strong relational embeddedness, but nevertheless they share little overlapping knowledge. As such, in an inter-organisational context, strong ties (i.e. channel members) are more likely to serve the bridging function through which information is transmitted than weak ties (i.e. competitors) because of their higher level of relational embeddedness and lower level of knowledge redundancy.

4.6 Conclusion

Weak ties are a key factor in the process of continuously reshaping existing rules and systems, and they are instrumental in helping different local projects to gradually add up to a technological trajectory at the global level. However, weak ties alone are insufficient to change socio-technical systems, because they hinder the transfer of complex knowledge and are accompanied by lower levels of trust. Knowledge of social network theory has thus proven valuable for providing insight into transition processes.

References

Bouty, I. (2000) Interpersonal and interaction influences on informal resource exchanges between R&D researchers across organizational boundaries. Academy of Management Journal 43 (1): 50-65.

Brown, J.J. and P.H. Reingen (1987) Social ties and word-of-mouth referral behaviour. Journal of Consumer Research 3 (December): 350-62.

Burt, R.S. (1987) Social contagion and innovation: cohesion versus structural equivalence. American Journal of Sociology 92: 1287-335.

Burt, R.S. (1992) Structural Holes. Harvard University Press, Cambridge, MA, USA.

Cross, R. and L. Sproull (2004) More than an answer: information relationships for actionable knowledge. Organization Science 15 (4): 446-62.

Dewar, R.D. and J.E. Dutton (1986) The adoption of radical and incremental innovations: an empirical analysis. Management Science 32 (11): 1422-33.

Geels, F. (2004) From sectoral systems of innovation to socio-technical systems: insights about dynamics and change from sociology and institutional theory. Research Policy 33: 897-920.

Geels, F. and R. Raven (2006) Non-linearity and expectations in niche-development trajectories: ups and downs in Dutch biogas development (1973-2003). Technology Analysis and Strategic Management 18 (3/4): 375-92.

Geels, F. and J. Schot (2007) Typology of sociotechnical transition pathways. Research Policy 36: 399-417.

Granovetter, M.S. (1973) The strength of weak ties. American Journal of Sociology 78 (6): 1360-80.

Granovetter, M.S. (1982) The strength of weak ties: a network theory revisited. In: P.V. Marsden and N. Lin (eds.). Social Structure and Network Analysis. Sage, Beverly Hills, CA, USA, pp: 201-233.

Hansen, M.T. (1999) The search-transfer problem: the role of weak ties in sharing knowledge across organization subunits. Administrative Science Quarterly 44 (1): 82-111.

Iansiti, M. and J. West (1997) Technology integration: turning great research into great products. Harvard Business Review 75 (3): 69-79.

Krackhardt, D. (1992) The strength of strong ties: the importance of philos in organizations. In: N. Nohria and R.G. Eccles (eds.). Networks and Organizations: Structure, Form, and Action. Harvard Business School Press, Cambridge, MA, USA, pp. 216-239.

Levin, D.Z. and R. Cross (2004) The strength of weak ties you can trust: the mediating role of trust in effective knowledge transfer. Management Science 50 (11): 1477-90.

Loorbach, D. and J. Rotmans (2006) Managing transitions for sustainable development. In: X. Olshoorn and A. Wieczorek (eds.). Understanding Industrial Transformation: views from different disciplines. Springer, Dordrecht, the Netherlands, pp. 187-206.

Madhaven, R. and R. Grover (1998) From embedded knowledge to embodied knowledge: new product development as knowledge management. Journal of Marketing 62 (4): 1-12.

Rindfleisch, A. and C. Moorman (2001) The acquisition and utilization of information in new product alliances: a strength-of-ties perspective. Journal of Marketing 65 (2): 1-18.

Rotmans, J., D. Loorbach and R. Van der Brugge (2005) Transitiemanagement en duurzame ontwikkeling: co-evolutionaire sturing in het licht van complexiteit. Beleidswetenschap 19 (2): 3-23.

Uzzi, B. (1997) Social structure and competition in interfirm networks: the paradox of embeddedness. Administrative Science Quarterly 42 (1): 35-67.

Wuyts, S., S. Dutta and S. Stremersch (2004) Portfolios of interfirm agreements in technology-intensive markets: consequences for innovation and profitability. Journal of Marketing 68 (2): 88-100.

Zucker, L.G., M.R. Darby and J.S. Armstrong (2002) Commercializing knowledge: university science, knowledge capture, and firm performance in biotechnology. Management Science 48 (1): 138-53.

Chapter 5

Doubt management as a tool for change: Karl E. Weick

Trond Selnes and Catrien Termeer

5.1 Maps of the unknown

> *The young lieutenant of a small Hungarian detachment in the Alps sent a reconnaissance unit into the icy wilderness. It began to snow immediately, snowed for 2 days and the unit did not return. The lieutenant suffered, fearing that he has despatched his own people to death. But on the third day the unit came back. Where had they been? How had they made their way? Yes, they said, we considered ourselves lost and waited for the end. And then one of us found a map in his pocket. That calmed us down. We pitched camp, lasted the snowstorm, and then with the map we discovered our bearings. And here we are. The lieutenant borrowed this remarkable map and had a good look at it. He discovered to his astonishment that it was not a map of the Alps, but a map of the Pyrenees. (Weick, 1995: 54)*

Weick uses this example to illustrate his concept of 'sensemaking':

> *Strategic plans are a lot like maps. They animate and orient people. Once people begin to act (enactment) they generate tangible outcomes (cues) in some (social) context, and this helps them discover (in retrospect) what is occurring (ongoing), what needs to be explained and what should be done (identity enhancement) next.... [T]he soldiers in this example were able to produce a good outcome from a bad map because they were active, they had a purpose and they had an image of where they were and where they were going. They kept moving, they kept noticing cues, and they kept updating their sense of where they were. (Weick, 1995: 55)*

Unfortunately, when people follow or create maps of an unknowable and unpredictable world, they face a strong temptation towards either overconfident knowing or overly cautious doubt (Weick, 2001: 361). In the first case, lack of sound doubt blocks out any progress of thought. In the second case, too much doubt inhibits progress. The soldiers nurture doubt in their continuous updates and search for cues. As such, they use organised doubt to make sense of uncertainty.

Transition management faces a situation similar to that of the soldiers. It addresses wicked problems by implementing deep long-term changes, the so-called 'transitions'. There is an almost inevitable sense of distance between the long-term goals and the short-term actions. By definition, future achievements are uncertain, while the short term calls for energising progress. Too much or too little doubt might disrupt this process of steered transition, leading to a sense of frustration. A focus on sensemaking could provide insights for dealing with this possible barrier. We use the concept of sensemaking here by applying the term 'doubt management'. The question is whether and how doubt management can help boost transition processes by empowering a course of action. We reflect on this question through the work of social psychologist Karl E. Weick and his concept of sensemaking.

Rather than providing a framework of analysis, this chapter aims to contribute to a mindset for dealing with transitions. It first introduces the term sensemaking, before dealing with some attributes of the relationship between sensemaking and transition management. It then moves on to suggest how doubt management can be applied for the purpose of transition management.

5.2 The concept of sensemaking

5.2.1 Sensemaking for sorting and connecting cues

Although Weick to our knowledge never addressed transition management in an explicit fashion, he surely engaged in tackling wicked problems and uncertainty. His appetite for subjects is large and varies from the dynamics of the Utrecht Jazz Orchestra to labour strikes in outer space (the Skylab crew) and to the Naskapi Indians of Labrador.

Despite his broad orientation, Weick is probably best known for his work on the concept of *sensemaking in organisations*. It started during his dissertation research in 1961, when he observed people who agreed to do a difficult concept-attainment task after learning that they would get less reward for their participation in the study than they had expected. Those who were deprived most severely rated the subsequent task as more interesting than the others. They were also three times more productive in performing it. Weick saw this as most striking. Why did they participate? The answer was simple: because the task was so interesting. But, said Weick (2001: ix), this answer is actually a complex mixture of 'prospective and retrospect sensemaking'. If action is public, irreversible and volitional, then a process of justifying the action is triggered. We like to think that what we are doing is worthwhile. However, in the process of making sense of things, people engage in fierce battles to create order out of life. People try to formulate and reformulate the past in order to win the interpretation game.

People expend a lot of energy creating order and clarifying preferences, means-goals relationships and cause-effect associations. Sensemaking is people's ongoing accomplishment of making retrospective sense of the situations in which they find themselves and their creations. Weick (1995: 49-50) distinguishes seven properties of sensemaking. It (1) is grounded in a continuing identity construction (2) based on perceptions of the world that are always retrospective (built on the past) and (3) can never be seen apart from the environment. Therefore, there is no such thing as a singular fixed environment detached from people. Instead, we constantly develop and negotiate the meaning of things. This is (4) a social process and (5) an ongoing process, without a real start or end. We judge our actions by the reaction of others and with (6) the use of extracted cues. Imagine that you are buying a piece of cloth and observe that 'it is going to fade'. If that judgement was made because you knew that the colour of the cloth was secured by a dye that was chemically unstable, and that this meant the colour would fade, then the notion of the dye, which is just one of many features of the cloth, is the extracted cue from which the character of the cloth itself is constructed. As we focus on extracted cues, we are (7) driven by plausibility rather than accuracy. We go for an explanation we can live with. We reduce paradoxes, dilemmas and inconceivable events to something we can comprehend. A search for accuracy, in certain forms, might even cause inertia.

We make sense of an activity or process by constructing, filtering, framing and creating meaning. This is a process of justifying action by looking back, listening to others and communicating with others. In so doing, we are creating 'meaningful lived experiences'. Groups form different stories based on different logics and cues. Weick refers to the influential distinction between threat and opportunity as contrasting labels for an experience. Viewing something as either a threat or an opportunity may dominate definitions of a project or situation and how we deal with it (Weick, 1995: 27). Because we have several projects, we engage in reflection. The problem is not that they make no sense, but rather that they make many different types of sense. Some of these might contradict others. In fact, Weick states, 'the problem is that there are too many meanings, not too few' (Weick, 1995: 27). The difficulty is often confusion, not ignorance. We do not need more information; we need values, priorities and clarity about preferences. But the time span between action and reflection is often short, 'which means that memory traces are typically fresh and rich with indeterminacy' (Weick, 1995: 29). The past is with us all the time, but we are 'mindful to just a handful of projects' (Weick, 1995: 29). In fact, we tend to settle for a feeling of order, clarity and rationality. Change is a challenge to this order, and that is probably why it is common to state that we resist change. Weick, though, sees this differently: 'The phrase "resistance to change", is organizational shorthand for the more general idea of "resistance to environment"' (Weick, 1995: 33). It does not necessarily mean combat against the environment. But we confront or await activities in the environment, and then we make sense of them, pretty much as the soldiers mentioned earlier.

'Sensemaking is tested to the extreme when people encounter an event whose occurrence is so implausible that they hesitate to report it for fear they will not be believed' (Weick, 1995: 1). Weick calls this the 'battered child syndrome', which was unknown until the 1950s. Prior to then, the problem 'did not exist'. In 1953, three cases were reported, in 1955 twelve, in 1967 more than seven thousand. By 1976, the number exceeded half a million. Attention exploded after the symptoms were given a label and publicly discussed. But our ability to label these symptoms was nonetheless slow to develop. Barriers remained to reporting them, as they still reflected 'hidden events'. Experts, for their part, overestimated the likelihood that they would surely know about the phenomenon if it actually were taking place. Doctors were unable to imagine that their previous diagnoses were all wrong, and there was no arena facilitating the creation of doubt. This is the 'fallacy of centrality': *because I don't know about this event, it cannot be going on*. It can discourage curiosity and create an antagonistic stance.

Weick states that if we extend this observation, it becomes conceivable that heavily networked organisations might find their dense connections unexpectedly risky, if they in their functioning encourage the fallacy of centrality. He uses the example of technology: the more advanced technology is thought to be, the more likely people are to discredit anything that does not come through it. But Weick also saw a striking difference in language between the phrase 'intentional ill treatment' and 'battered child'. The first is rather neutral, while the second can mobilise outrage and action (Weick, 1995: 4). The more general point, Weick states, is that 'vivid words draw attention to new possibilities, suggesting that organizations with access to more varied images will engage in sensemaking that is more adaptive than will organizations with more limited vocabularies' (Weick, 1995: 4).

In the process of sorting through prior cues, labelling them and connecting them, we end up with plausible stories that are good enough to keep us going. The stories culminate in meanings based upon plausibility rather than accuracy, and numerous features are overlooked and forgotten along the way. This is also central to Weicks' understanding of learning: 'The process of learning occurs when people notice some of what was previously overlooked and overlook some of what was previously noticed' (Weick, 2001: 305).

Both aspects are demanding. Pushing something back into a story as well as pulling something out of a story requires resources. We make sense through an ongoing concentration on a few critical problems, learning the history of them, building coalitions and mobilising support, as best as we can. In essence, this is cryptic, incomplete and tentative, and it depends on how we organise, as Weick illustrates in this example:

> *If you place bees and flies in a bottle and lay it down on the table with its base towards the window, you will find that the bees will persist on seeking a way towards the base, even until they die. The flies on the other hand will sally forth and within two minutes be out of the bottle. Why is that? Why do the flies, with their feather-brained lack of logic, solve something the far more intelligent bees cannot cope with at all?*
>
> *For the bees it is their love for light, their very intelligence, that is their undoing. They act in accordance with their logic; escape is where the light is clearest. Glass is for them a supernatural mystery, it is an inadmissible piece of information,*

incomprehensive. The flies are also ignorant to glass, but also equally careless to the logic of light. They flutter hither and thither, and meeting the good fortune that often waits on the simple, they will necessarily end by discovering the friendly opening that restores their liberty to them. (Weick, 2001: 381).

The episode above becomes disastrous for the bees because they had the solution to the problem while being unable to make sense of the problem. Their structure of sensemaking is too tight and rigid. There is already 'one best way of action'. The flies go for randomness and trial and error, allowing detours and moves back and forth. There is a striking difference between these two worlds: they differ in the degree to which means are tied to ends, action is controlled by intentions and rigid logic dominates exploration. In this case, looseness is an asset and tightness is not. Weick argues that this need not always be the case, but it is likely if the environment is very diverse. The point is that organisations vary in the loose-tight dimension. For Weick (2001: 383), loosely coupled elements exist when A affects B,

- suddenly rather than continuously;
- occasionally rather than constantly;
- negligibly rather than significantly;
- indirectly rather than directly;
- eventually rather than immediately.

Loosely coupled organisations have delays and lags; they also might be unpredictable and erratic, with untrustworthy feedback (Weick, 2001: 384). Sensemaking within these organisations is likely to be better suited to complex environments. Referring to Asby's law of requisite variety (*only variety can beat variety*), Weick (1995) argues that it takes a complex sensing system to register a complex environment. A danger here might be an overly loose and fragmented organisation, which has no direction and an escalating problem of coordination (Weick, 1995). But a loosely coupled organisation has the potential to be a generalist-oriented self-organising type with great ability to adapt. Moreover, due to their flexible learning ability, very loosely coupled organisations may be able to remain rather stable.

A tight structure might be good for bringing single products into a market. The chain of command is then known and rule-based, with predictability and a rigid system of identifying errors and organising feedback. But rigidity, though suitable for a steady course, might be a hindrance to change, as the sensemaking might tend to simplify and categorise a complex environment, (pre)structuring the possible actions and preventing novelty. However, even tight organisations, such as large public bureaucracies with little chance to adapt to rapid environmental change, are inhabited by people who are able to engage in adjustments to new environments (Termeer and Van der Peet, 2009). Very tightly coupled organisations are able to maintain stability based on their own structure and chain of command.

Sensemaking also varies according to the type of change at hand. Weick and colleagues refer to two opposite types: episodic change and continuous change (Weick and Quinn,

1999; Termeer and Van der Peet, 2009). 'Episodic change' is a fundamental shift in values, structures, procedures, beliefs and activities during a short period. 'Continuous change' is an ongoing process of adjustment to, or experimenting with, contingencies, exceptions, opportunities or unintended consequences. It is more of a constant update, but shifts in practice also create new conditions for breakdowns and innovations. Change is emergent and new patterns of organisations occur without *a priori* intention (Weick and Quinn, 1999). But small, continuous adjustments are not trivial, as they can accumulate, amplify and create huge shifts. Loosely and tightly coupled organisations react differently to these types of change. We associate loosely coupled organisations with a stronger ability to deal with episodic change. But both types of organisation are probably too inert to change as rapidly as the environment (Weick and Quinn, 1999). New demands too might become threats for both because they unintentionally mobilise defensive routines, causing stagnation of learning (Argyris, 1999). To overcome stagnation, an organisation must loosen fixations, restructure and try to change meaning systems and build commitment. Tightly coupled organisations might find this more challenging than loosely coupled organisations.

5.3 Making sense of transitions

This section discusses some attributes of the relationship between sensemaking and transitions. In particular, it focuses on efforts to introduce transition management as a tool for dealing with wicked problems. At that stage, creating a transition arena for mutual agenda setting is crucial. This idea might sound refreshing and interesting, but it is also an unusual concept. *It has no past.* There are no images of transitions in our perception. The sensemaking has thus 'just started' and we need to be convinced, which makes our trust in the concept as yet undetermined. This section discusses four aspects of this sensemaking process:
- the transition roll call to trigger interest;
- the danger of alienation;
- tolerance as a way out (less hubris and more fun!);
- the pitfalls of tolerance (the delicacy of 'solutions we can live with').

5.3.1 The transition roll call to trigger interest

One attribute of the 'call to transition' is its strong ability to trigger attention and interest by its focus on joint problem definitions and agenda development, directed towards experimentation and learning. It is suited to signal an urge to change by including people and connecting them to wicked problems. In that way it might energise people where frustration used to prevail. It clearly breaks with more conventional types of management, and as such it serves to avoid a sense of 'business as usual'. The novelty then attracts the attention of a wide range of participants. If for no other reason, people's curiosity is aroused. They also meet each other in a new setting, allowing them to see new opportunities and maybe to define old problems in novel ways. In so doing, transition management might change the way people make sense out of wicked problems. This makes it a welcome addition to existing modes of working. The initial attraction, however, is soon swept away if participants are alienated from the very process of change itself.

5.3.2 The danger of alienation

The attractiveness of a concept promising and proclaiming change is great, and it teases and triggers people to act upon it. The challenge, however, is to place the call to transition within a workable framework, a framework that provides a trustworthy promise of progress. Advocates of transitions are very much aware of this, as when Rotmans (2003: 45) warns of dangers such as backlashes and lock-ins due to inadequacies in design, too fast departures or an embedding that renders change impossible. But, as stated by Rotmans *et al.* (2005: 20), while transition concepts are still at an early stage of development they need to prove themselves in practice. One could be 'in possession' of concepts derived from experience, maybe even concepts that have been tested in reality; yet, notions like 'transition paths', 'S-curves' and 'co-evolution' are probably not common in the language of possible participants. Their high potential might come with correspondingly high expectations. If practice proves disappointing, people easily distance themselves from the language. Or, if motivation is high, the transition approach will probably be allowed some time to prove itself worthwhile (but not very long).

Rotmans *et al.* (2005: 11) build bridges from concept to practice using a whole set of operational notions regarding programmatic learning, building joint agendas and arenas, and networked solutions. Despite the best of intentions, it might prove quite demanding to step from these wonderful sounding terms of novelty into the possibly more down-to-earth matters of what to do in actual practice. The stretch of participants' enthusiasm is likely to be seriously tested if the exciting prospect of being part of an agenda-building transition arena plummets into a feeling of having landed in yet another incomprehensible meeting. Also, the transition manager might feel ill at ease under the heavy burden of the new concepts without having some practical equivalence. Paralysing doubt may then turn the conceptual beauty into a sense of ugliness. The potential goes unrecognised in practice, and the danger of alienation from the whole process becomes overwhelming.

5.3.3 Weick's call for tolerance: less hubris and more fun!

Weick (2001) proposes dealing with uncertainty and change by developing a tolerance for ups and downs by accepting these as normal fluctuations rather than viewing them as a testimonial to one's own shortcomings. Weick (2001: xi) suggests that we should take ourselves a little less seriously: a little less hubris and a little more fun! Weick's work often deals with fixations on certain values and norms, where a sound dose of doubt is often lacking or overabundant. Transition management has the potential to state that 'what we have done so far is not worthwhile' (it is not sustainable), but at the same time it offers an energising way out, through the transition paths and arenas. Weick's call for tolerance and fun is a way of avoiding alienation and a sense of 'mission impossible'. Transitions, then, should become serious fun.

5.3.4 Pitfalls of tolerance: the delicacy of 'solutions we can live with'

Tolerance to normal fluctuations, as suggested above, contains specific pitfalls when viewed together with our inclination to create plausible solutions that we can live with. Instead, we should be better equipped to address how problems get defined in the first place. Focusing on problems is of vital concern but also too general for any practical purpose. The cue to progress is awakening genuine interest. 'Ironically, people often can't solve problems unless they think they aren't problems', says Weick (2001: 427). We often define social problems in ways that overwhelm our abilities to do anything about them (Weick, 2001: 426). We easily settle for solutions we do not like simply because at some point in time they are solutions we can live with, especially if they are the outcome of our established ways of dealing with problems. Genuine interest is then needed to break out of this pattern.

5.4 Doubt management as action

This final section explores options for employing doubt as an asset rather than a liability. The idea is that it may contribute to practical transition management. We make six suggestions:
- name and rename doubt and tensions;
- challenge causal expectations (or, hunt a caribou);
- drop the heavy tools of rationality, and learn to say 'I don't know';
- move from ignorance to decisive action;
- appreciate the great role of small wins;
- encourage shared experiences.

5.4.1 Name and rename doubt and tensions

Managing doubt is a matter of crafting expectations. People need to access what they expect in important situations and then deliberately look for disconfirming data (Weick, 1995: 190). An essential management task is to somehow balance the long-term complex uncertainty and the short-term plan-oriented means-goals certainty. The sensemaking might fall short if statements and clichés outshine sound doubt. It is up to management to provide plenty of subtlety and nuance. Otherwise, the intended innovation is shut down (Weick, 1995: 183). Explaining and discussing basic concepts are vital components of this process, coupling the long-term and short-term mission. The core business of transition management is to spend ample time naming and renaming the doubt and allowing the corresponding tensions some air. Absence of tension is worrisome, as few changes are likely to occur without some emotion and (hopefully creative) friction. An absence of tension might signal a lack of genuine interest and sense of urgency. To trigger tension and genuine interest, management must challenge participants' causal understanding.

5.4.2 Challenge causal expectations (or, hunt a caribou)

An example of managing the unexpected used by Weick is the hunt for caribou by the Naskapi Indians of Labrador. 'Their problem is where to hunt for caribou. The hunters hold the shoulder blade of a caribou over a fire until it develops cracks. Then somebody reads those cracks to see where the caribou are likely to be' (Weick cited in Geirland, 1996). The wisdom of this practice, Weick states, is that it randomises the hunter's behaviour, making it harder for the caribou to learn where the hunter is likely to be. It also ensures that areas do not become overhunted. The reader of the bones is crucial. Past experience is given some weight, as the reader injects some experience into an interpretation of what the cracks mean. 'If the reader's hunches dominate, randomization is lost. If the cracks dominate, experience is discarded' (Weick, 2001: 113). The obvious is thus not always the best way. From our western point of view this method might seem inadequate based on our notion of 'proper' cause-effect analysis. But our established cause-effect arrangements are often the problem in the first place.

Expectations must relate to the unexpected, and this is the failure of many management schemes. Expectations are built into roles, routines and strategies (Weick and Sutcliffe, 2007: 23). Expectations create blind spots, Weick and Sutcliffe emphasise. These authors (Weick and Sutcliffe, 2007) stress that building capabilities to enrich awareness is vital for organising doubt. People committed to the task of transition must be able to thrive in an error-friendly 'learning culture'. This is about grasping inertia, establishing insight into 'near misses' and 'good news', and training people to raise new information and cast doubt but not blockades, making it normal to treat unexpected information as new information. Making sense of doubt also entails a process in which the core members spend time on the front line of events, avoiding a division of labour by which some give orders to others without knowing how it works. Giving orders is not a useful activity in itself; the order lacks a sense of meaning.

Weick and Sutcliffe (2007: 156) refer to a practical tool to make sense of decisions, developed by Gary Klein and labelled 'STICC'. According to this tool a briefing should involve the following elements:
• the *situation* ('here is what I think we are facing');
• the *task* ('here is what I think we should do');
• the *intent* (here is why I think that is what we should do');
• *concerns* ('here is what we should keep our eye on because if that changes, we're in a whole new situation');
• *calibration* ('tell me if you don't understand, cannot do it, or see something I do not').

However, people are reluctant to say 'I don't know'.

5.4.3 Drop the heavy tools of rationality, and learn to say 'I don't know'

Should we engage more in the capability to say 'I don't know'? What happens when the manager says 'I don't know'? According to Weick (2009: 268), such a manager is essentially saying that the group is facing a new ball game. It is time 'to drop the heavy tools of rationality' and in so doing 'gain access to lightness in the form of intuitions, feelings, stories, experience, active listening and…empathy'. There is not as much softness here as one might expect. A manager who simply says 'I don't know' is up for a hard challenge, as conventional wisdom demands a clear answer involving how to understand. But in fact, 'we know little and understand even less' of this modern and complex society, to borrow the words of Witte *et al.* (1990: 17). In 1985, the University of Arizona College of Medicine even established a 'house of ignorance' to come more to terms with what we do not know. It aimed to help students recognise and deal with the vast intimidating world of non-knowledge, composed of the things we know we don't know, the things we don't know we don't know, and the things we think we know but don't. This decade's facts might become the next's follies and vice versa. We need skills to work with this, to understand uncertainty and ambiguity. To learn these, students would be graded not by their short-answer tests but by the progression of their questions.

5.4.4 Move from ignorance to decisive action

The move from acceptance of ignorance to decisive action is the next step. This is the part where people discover that they have been decisive all along. But, due to the carefully performed process of naming and renaming doubt, challenging the long frozen causal establishment and dropping the heavy tools of rationality, we now move to a type of decisiveness where the manager no longer is accused of hypocrisy or vagueness every step of the way. At least, this risk is less prominent as managers become better suited to explaining what they know and do not know, and also what others know and not know. According to Weick (1995), 'Even if people exaggerate the similarity between the streamlined past and the disorderly present, this very exaggeration may make them more confident' (Weick, 1995: 184). As those involved in the transition challenge know more about the past and the present, they also empower each other. 'Confident people are more likely to put in place the environment they expect and can deal with' (Weick, 1995: 184). They are better equipped to facilitate the transition by creating expectations and scripts that function like self-fulfilling prophecies. But to enable progress, there is at least one more facility to consider. That is, the great role of small wins.

5.4.5 The great role of small wins

Although most people have intense feelings about social problems, interest in any given issue soon diminishes without results, and people wander off bored (Weick, 2001: 440). A sound transition scheme, therefore, must be aware of the small wins. Small wins refer to the identification and display of minor successes on the long road to sustainability. The challenge is to orchestrate a process whereby taking distance from the original problem definition is central. Weick calls this the tactic of 'discrediting'; purposely turning one's back to the past. Weick argues that responses that are more complex, more recently learned, and more responsive to more stimuli in changing situations usually have a better chance of producing lasting changes in perceptions of dynamic problems (Weick, 2001). Small wins might even engineer great success. But small failures must also be scrutinised (Weick, 2009). In so doing, much needed motivated critical voices are added. But there is no substitute for success at the end of the day.

5.4.6 Encourage shared experience

We tend to label problems to fit the solutions we are engineered to make. In other words, we couple problems to pre-selected solutions. Transition management is about breaking that pattern. It demands cross-sector inter-disciplinary collaboration. This is no minor challenge, as it wakes up all kinds of sceptical and allergic reactions. Although shared meanings might be difficult to attain, they are the glue of a group. They can also be organised. A joint programmatic approach to create shared experiences is then no luxury. By that, a shared language of expectations is created. Expectations, Weick argues, are real enough, but they need to be crafted with care (Weick, 1995: 189-190).

5.5 Conclusion

We have argued here that doubt management could add to the quality of transition management. We need to break out of organisational cultures that tend to treat the existence of multiple meanings and interpretations as a symptom of weakness rather than as an accurate barometer of turbulence. Too often we face too many cues, too many interpretations and too little closure (Weick, 1995: 186). Things must make sense to us, if we are to stay in the game. And to animate and orient, vigour is better than rigour, as in the example of the soldiers.

References

Argyris, C. (1999) On Organizational Learning. Blackwell Publishing, Boston, MA, USA.

Geirland, J. (1996) Complicate yourself. Interview with Karl E. Weick, Wired (April). Available at: www.wired. com/wired/archive/4.04/weick_pr.html.

Rotmans, J. (2003) Transitiemanagement, sleutel voor een duurzame samenleving. Van Gorcum, Assen, the Netherlands.

Rotmans, J., D. Loorbach and R. Van der Brugge (2005) Transitiemanagement en duurzame ontwikkeling: co-evolutionaire sturing in het licht van complexiteit. Beleidswetenschap 19 (2): 3-23.

Termeer, C.J.A.M. and G.F.V. Van der Peet (2009) Governmental strategies and sustainable transitions: monitoring systems for the prevention of animal disease. In: K.J. Poppe, Termeer, C., Slingerland, M. (eds.). Transitions towards sustainable agriculture and food chains in peri-urban areas. Wageningen Academic Publishers, Wageningen, the Netherlands, pp. 253-273.

Witte, M.H., C.L. Witte and D.L. Way (1990) Medical ignorance, AIDS-Kaposi's sarcoma complex, and the lymphatic system. Western Journal of Medicine 153 (July): 17-23.

Weick, K.E. (1995) Sensemaking in Organizations. Sage Publication, Thousand Oaks, CA, USA.

Weick, K.E. (2001) Making Sense of the Organization. Blackwell Publishing, Oxford, UK.

Weick, K.E. (2009) Making Sense of the Organization: the impermanent organization (Volume 2). John Wiley and Sons Ltd., Hoboken, NJ, USA.

Weick, K.E. and R.E. Quinn (1999) Organizational change and development. Annual Review of Psychology 50: 361-86.

Weick, K.E. and K.M. Sutcliffe (2007) Managing the Unexpected. Second edition, John Wiley & Sons Inc., Hoboken, NJ, USA.

Chapter 6

Power and discipline in transitions: Michel Foucault

Martijn Duineveld and Guus Dix

Martijn Duineveld:	*Est-ce qu'on peut dire que les voleurs dans les Ministères, essayant de trouver des papiers sur la construction des usines nucléaires, font, a votre avis, part d'une transition vers une Hollande plus durable, disons qu'on considère l' énergie nucléaire comme durable?*
Michel Foucault:	*Bien sur!*
Martijn Duineveld:	*Mais pourquoi est-ce qu'on ne lit pas ça dans la littérature sur les transitions?*
Michel Foucault:	*Une belle question a laquelle on peut répondre plusieurs choses. Disons premièrement que...*

6.1 Introduction

How wonderful it would be to have the philosopher most cited within the social sciences, Michel Foucault, shed light on the still largely Dutch phenomenon of 'transition management'. This would undoubtedly provide us with an exquisite example of intellectual fireworks. Sadly, Foucault died of AIDS in 1984. So we have no choice but to do the job ourselves. In order to do so we will link our interpretation of his work with the field of transition management (Kemp *et al.*, 2007; Duineveld *et al.*, 2009). We bypass his early (Foucault, 2001, 1973, 1969) and late work (Foucault, 1985, 1986) and base this chapter primarily on Foucault (2004, 2003, 1998 and 1994a). We also use secondary literature like Flyvbjerg and Richardson (2002), Flyvbjerg (1998a), McHoul and Grace, (1995), Van Assche (2004) and Gutting (1994, 1989). These texts are used pragmatically; that is to say, we do not attempt to make a connection with the schools of thought that were of influence on Foucault's thinking, such as Marxism, structuralism and analytical philosophy. Neither do we enter into the (philosophical) movements he opposed, such as phenomenology and existentialism. Instead, we use his texts as a toolkit, as an initial impetus for deriving a number of theories, concepts and ideas that can help us to describe, analyse and understand transition practices.

This chapter should be read as an exploration of Foucault's thinking for the analysis of political, administrative and social transitions. In it, we first describe Foucault's ideas on the entanglement of power and knowledge: 'power/knowledge'. Then we describe three mechanisms: disciplining, subjection and exclusion. These mechanisms make the power/knowledge entanglement explicit. Each mechanism is related to transition management by means of an example. Finally, we argue that looking at transitions with Foucault in mind shows us how these can be analysed, both in a cynical and in a reflexive way.

6.2 Foucault

6.2.1 Power/knowledge

What I would like to tell you in these lectures are some things that may be inexact, untrue, or erroneous, which I will present as working hypotheses, with a view to a future work. I beg your indulgence, and more than that, your malice. (Foucault, 1994a: 1)

These are the first sentences of a series of five lectures Michel Foucault delivered at a Catholic university in Rio de Janeiro in 1973. The Catholic intellectuals attending the lectures certainly must have thoroughly enjoyed themselves. Foucault reminded them that God had been declared dead by Nietzsche at the end of the 19th century, and he tore up some of the premises that had been fundamental to Western philosophy since Plato. He also clearly described his use of the concept of discourse.

For Foucault, discourse is a practice within which realities are produced. The elements that constitute a discourse can be considered games. He approached these as strategic games of action and reaction, question and answer, domination and evasion, as well as struggle. At one level, discourse is a regular set of linguistic facts. At another level, it is an ordered set of polemical and strategic facts (Foucault, 1994a). Linguistic expressions are not necessarily central in Foucault's approach; rather, they are considered part of the many practices that make up a certain discourse. Foucault makes no ontological distinction between the linguistic and behavioural aspects of the practice either (cf. Howarth, 2000; Laclau and Mouffe, 1985). A discourse is made up both of linguistic elements and of all other conducts and signs that refer to it (Hall, 1997). The pronunciation and writing of words, an anti-war demonstration, the publishing of a scientific book; all of these activities can be regarded as practices taking place within a discourse, and simultaneously giving shape to that same discourse.

Foucault opposes the idea that knowledge, truths, facts or rationality should be considered as mirrors of reality. According to Foucault, it is not pure reason that determines what people hold to be just and true. Nor does the world *itself* determine how it is perceived, in what manner people in scientific and non-scientific practices discuss it, and how it is dealt with. Rather, it is the practices in which people, organisations and institutions operate and the society in which they move that collectively produce a certain reality, a certain discourse. These powers constitute that the statements certain people make, within a certain discourse, are accepted as truths, while those of others are dismissed. To Foucault, knowledge, truth and rationality are thus contingent and arbitrary. Contemporary academics would now say that they are social constructions, produced within social practices.

According to Foucault, knowledge, rationalities and truths mutually presuppose and define each other. Knowledge is power, and power also defines what is considered as valid knowledge, as truth or as reality within a certain discourse. Inspired by Foucault, the planning scholar Flyvbjerg illustrated this mechanism well (Flyvbjerg, 1998b). In his book *Rationality and Power: Democracy in Practice*, a study of planning practices in the Danish city Aalborg, he describes how various actors involved in the planning process, often with conflicting interests, very selectively produce particular knowledge. In order to increase their influence on the planning process, they make sure that the knowledge that is not to their advantage is concealed, marginalised or buried under the proverbial pile of paper. Flyvbjerg sums up his project as follows:

> *I already mentioned...Francis Bacon's dictum that knowledge is power. This dictum expresses the essence of Enlightenment thinking. 'Enlightenment is power' and the more enlightenment – the more rationality – the better. The Aalborg study shows that Bacon is right; knowledge is power. But the study also shows that the inverse relation between power and knowledge holds and that empirically, as opposed to normatively, it is more important: 'Power is knowledge.' In this sense, the study stands Bacon on his head. It shows how power defines what gets to count as knowledge. It shows, furthermore, how power defines not only a certain conception of reality. It is not just the social construction of rationality that is at issue here; it is also the fact that power defines physical, economic, social, and environmental reality itself. (Flyvbjerg, 2002: 361).*

For a proper understanding of Foucault's ideas on power/knowledge, the concept of power needs to be explicated further, since he uses it differently from its use in everyday language. Foucault's conception of power can be understood in contrast with a few common lines of thought which hold that: power is negative; power is repressive; power is mostly exercised by means of laws; power is that which is exercised over people through the judicial system; power expresses itself in rules; power is a possession; power is owned by a sovereign or by the state, by the system; proper knowledge is developed independently from power; power is visible, and so on.

For a long time, these conceptions structured thinking about power and power relations, and this is still the case within certain political and scientific discourses, albeit implicitly (Foucault, 1998). Foucault rigorously dissociates himself from these conceptions of power:

> *[There] are readymade models: when one speaks of power, people immediately think of a political structure, a government, a dominant social class, the master and the slave, and so on. I am not thinking of this at all when I speak of relations of power. I mean that in human relationships, whether they involve verbal communication ..., or amorous, institutional, or economic relationships, power is always present: I mean a relationship in which one person tries to control the conduct of the other. So I am speaking of relations that exist at different levels,*

> *in different forms; these power relations are mobile, they can be modified, they are not fixed once and for all. (Foucault, 1997: 292)*

Power should therefore be seen and studied as a mechanism that is operative everywhere and that is exerted from various positions (Foucault, 1998: 93). Moreover, power relations are always intentional; they are exercised with a certain aim. Foucault states that there is no power that is exercised without a series of aims and objectives (Foucault, 1998: 93). Here the word power does not have a negative connotation, contrary to its everyday use. Power is neither good nor evil. It can have both oppressive and creative consequences. It produces some discourses, realities, knowledge and values and pushes others into the background (Foucault, 1998, cf. 1994c).

These conceptualisations of power/knowledge are made explicit next when we elaborate on three interconnected mechanisms: disciplining, subjection and exclusion.

6.3 Disciplining, subjection and exclusion

We now derive a number of mechanisms from Foucault's work that could be of great value for the analysis of transitions. Without presuming these are present in every transition per se, the following mechanisms come to the fore when studying transitions from a Foucauldian perspective.

6.3.1 Disciplining

Foucault made a detailed study of the relation between knowledge and power in his book *Discipline and Punish* (Foucault, 1979). The book can be seen as an analysis of the transition of forms of punishment. Some people have characterised this transition as a process of civilisation, in which the pre-modern barbarity of the Middle Ages (executioners, scaffolds and the like) were transformed into the present-day, humane prison system. Foucault provides us with a different reading. The transition from corporal punishment to imprisonment should not be seen as a process of modernisation and humanisation, but as a complete reorganisation of the way power is exercised in our penal system. He argues that the public nature of the 'technologies of punishment' was an essential part of the monarch's power. Through these spectacles of torture and suffering, the sovereign showed strength to its subjects. At the same time, however, the public character of these penal practices was also the weak link of the power ritual. Spectators at the executions cheered, but they protested as well. Sometimes they declared themselves openly in favour of the condemned men and women or as against certain executions and thus against the sovereign. At the end of the 18th century a whole generation of penal reformers emerged that put these weaknesses of the sovereign exercise of power at centre stage. Were people getting used to the cruelties that they should be disgusted about? And why was power clenched on the few instances that it was present, while all kinds of illegal conduct went unnoticed or were even considered common practice? What is thus often described as the rise of a modern penal discourse in which 'man' and the 'humane' were introduced in a plea for a more 'gentle' criminal justice system, is according to Foucault a general critique of a bad 'economy of power'.

These reformers imagined a new range of punishments that were more precise, effective and encompassing. It was this range that Foucault tried to capture with the notion of 'discipline'. In contrast to earlier forms of punishment, disciplinary techniques attempt to enforce people's conformity to certain standards of behaviour without physically harming them. The function of punishment thus became that of a correcting mechanism, from the timetables specifying what inmates have to do at certain times of the day to the complete configuration of space such that inmates are always visible to those supervising them. According to Foucault, disciplinary techniques are not restricted to a specific institution, like the prison. They are widespread in our society. From cradle to grave, we move through maternity and infant centres, educational institutions, healthcare checks and networks of security cameras that 'keep watch' over public and private spaces (cf. Koskela, 2000).

The example of prison life might seem somewhat remote from the practice of, for example, the transition towards a sustainable society. Nevertheless, disciplinary mechanisms that try to mould behaviour in conformity with pre-established objectives can be found there as well. The introduction of environmental regulations, for example, can have an effect on the distribution of means within the agricultural sector. In a similar vein, control systems regulate the conduct of agrarians so that they become 'environmentally friendly'. Agrarians and their farms are under constant supervision, and the requirements change from time to time.

6.3.2 Subjection

Another mechanism deductible from *Discipline and Punish* is that of subjection or *assujettissement* in French. Both terms capture the double meaning of the word 'subject'. On one hand, a subject is an individual subjugated to another individual in a relation of dependence and control, as is the case when a ruler rules over subjects. On the other hand, a subject is an individual that is tied to a certain identity it considers its own. Combining these two aspects, subjection is thus for Foucault a form of power directed towards an individual that 'marks him by his own individuality' (Foucault, 1994b).

Over and against the disciplinary rules and regulations, subjection is both a 'softer' as well as a more 'in depth' technique of power, as it concerns both the creation of novel forms of subjectivity and the way people affirm or resist being categorised thus. Contrary to the power/knowledge mechanism of discipline, subjection is not only directed at specific behaviour of, say, delinquents, soldiers and students, but also constructs them as subjects. Thus, Foucault argues that these 'subjects' are not simply a natural, biological reality; they are social constructions.

The production of these subjects or species is not the same thing as the development of typologies for naming new phenomena. Consider delinquency. In the good old days of corporal punishments, the question as to the nature of the offender was inexistent. With the rise of 'humane' measures aimed at correction, the judicial apparatus began to look for the hidden causes that made offenders do the things they do. The figure of the 'delinquent' now appears as someone inherently disposed to all sorts of criminal behaviour, whether because of 'biological deficiencies' or a 'bad youth'. Or consider homosexuality. It wasn't like sex between men was a new phenomenon and then people named it. The practice existed before that. It is rather the case that for these kinds of sexual practices there was no discourse to name them, to distinguish between them and to place them in a normative hierarchy. 'Normal sexuality' between husband and wife was set apart from 'abnormal sexuality', for instance, between homosexuals. A man who slept with other men once in a while became 'gay' and was part of the gay community: 'The homosexual was now a species. So too were all those minor perverts' (Foucault, 1998: 43; cf. Hacking, 1999).

The same type of question that Foucault introduces when speaking about the emergence of 'delinquency' and 'homosexuality' can be posed with regard to sustainability. Of course, rules and regulations go a long way. But what if we can bring people to identify with the ideals we try to impose on them? How can they be led to recognise themselves as 'sustainable subjects' and act accordingly? There are undoubtedly many (unintentional) mechanisms of subjection within the transitions aimed at a sustainable society. Within transition discourses identities change and new subjects emerge. In the last 40 years, for example, environmentalists have been subjected into 'alternative people', 'progressive people', 'radicals', 'extremists', 'softies', 'nimbys' 'establishment', 'forerunners' and 'innovators', depending on the fashionability of environmentalism. Thereby they have gained or lost power. The same goes for their enemies, the 'selfish people', the 'sceptics', the 'neo-liberals and the 'destroyers of planet earth'.

6.3.3 Exclusion

On 2 December 1970, Foucault gave his inaugural speech *L'ordre du discours* for the prestigious College de France (Foucault, 2004). In this speech he argued that various exclusionary mechanisms are active in our society. These mechanisms simultaneously control, select, organise and redistribute the production of the discourse, both from the inside and from the outside. Exclusion can mean all power mechanisms that place people, ideas or knowledge outside of certain discourses and practices.

Exclusionary mechanisms are inextricably bound up with the production and regulation of a discourse within a network of social practices. Take the ban, for example, by which we do not have the right to speak about everything under all circumstances. Certain subjects are taboo; and certain speakers have exclusive rights or privileges. Exclusion can be made up of mechanisms that are purposefully used within a certain discourse to continue power relations, as well as mechanisms that are unintentionally incorporated in particular practices, systems or instruments. In short, by exclusion we understand all mechanisms that cause people and the organisations they are in to exclude the thinking, acting and speaking of other groups and organisations, whether they intend to or not.

Particularly the exclusion of naïve, marginal and disqualified knowledge gets a lot of attention in Foucault's work. In his historical research he compared the knowledge that was disclosed in certain periods with the dominant discourses of that same period. This was not just to show that there were marginal discourses alongside the prevailing ones, but also because their simultaneous existence reveals how the dominant discourses managed to produce and reproduce themselves by marginalising particular knowledge and thus excluding it. Foucault said that within certain discourses some knowledge has been qualified as 'nonconceptual knowledges, as insufficiently elaborated knowledges: naive knowledges, hierarchically inferior knowledges, knowledges that are below the required level of erudition or scientificity' (Foucault, 2003: 7). It is not that knowledge is hidden or tucked away; rather it is placed in a hierarchical position in relation to other types of knowledge: 'They are located low down on most official hierarchies of ideas…. Certainly they are ranked "beneath" science. They are the discourses of the madman, the delinquent, the pervert and other persons' (Foucault in McHoul and Grace, 1995: 15).

Exclusionary mechanisms are an aspect of both discipline and subjection. When certain kinds of behaviour and subjectivity are promoted, other forms become redundant or inferior. An important argument is that inclusionary and exclusionary mechanisms are part of every social change or transition. Evidently, defending the necessity of a 'transition' implies the subordination of certain current practices. What do these practices mean to those engaged in them? In what way do they accept or resist being an element in a large-scale transition? In line with Foucault, Scott provides many examples of exclusionary mechanisms in transitions inspired by the modernist project (Scott, 1998). The title of his book provides a good summary of its content: *Seeing Like a State: How Certain Schemes to Improve the Human Condition Have Failed*. Scott argues that the modernistic transitions to improve agriculture, forestry and city planning have had devastating side effects due to the quest for control and legibility, which have led to simplification and the exclusion of subtle local knowledge and practices.

6.4 Michel Foucault and the transition researcher or manager

To what extent does Foucault's analysis provide clear-cut prescriptions for the fields he scrutinised? What do Foucault's insights mean for those studying and managing transitions? The answer is not straightforward. On one hand there are hardly any prescriptive passages in his works. Foucault believed he should only be descriptive, not prescriptive when it concerned political questions:

> *And no doubt fundamentally it concerns my way of approaching political questions. It is true that my attitude isn't a result of the form of critique that claims to be a methodical examination in order to reject all possible solutions*

except for the one valid one. It is more on the order of 'problematization' – which is to say, the development of a domain of acts, practices, and thoughts that seem to me to pose problems for politics.... [For] crime and punishment...it would be wrong to imagine that politics have nothing to do with the prevention and punishment of crime, and therefore nothing to do with a certain number of elements that modify its form, its meaning, its frequency; but it would be just as wrong to think that there is a political formula likely to resolve the question of crime and put an end to it. (Foucault, 1997: 114)

Following Foucault, here one could conclude it is not the role of academics to produce political formulas, management rules and tactics which inform people how to think, speak and act. Instead one should analyse the consequences of certain practices of thought, speech and acting. Metaphorically speaking, Foucault did not propose plans for renovating buildings; rather he provided detailed analyses of the state they were in.

On the other hand, Foucault was also politically very active. He produced his works in close relation to his own life experiences and political activism. For example, *Discipline and Punish* was published at a time when Foucault was involved in the Prison Information Group, which had a critical influence on the French penal system. He used his own writings and his status as expert on the subject matter to legitimise and underpin his own political actions. So, although Foucault aimed to be as analytical and reflexive as possible, his works were used and played a role in many contemporary political discourses.

Taking into account these seemingly contrary ways of dealing with the relation between description and prescription, we see two possible uses of Foucault's work for the analysis and management of transitions: a cynical one and a reflexive one. The cynical use of Foucault constitutes a deliberate attempt to uncritically exercise his insights, like the ones on disciplining, subjecting and exclusion (as well as marginalisation, subjugation, control, et cetera) as means to achieve the goals in a certain transition process. We call this the cynical use because it employs Foucault's critical accounts of power/knowledge mechanisms in an uncritical, non-reflexive, pragmatic and even opportunistic way to achieve certain goals. It is like reading and applying Machiavelli outside of its historical context.

We feel less discomfort in using Foucault as a form of critical, reflexive analysis to reveal the hidden suppositions and the problems and unintended consequences of policies. Foucault's thinking on power/knowledge can be used as a way to continuously criticise the thinking and acting of people and institutions involved in a transition process, including the researchers. The reflexive and critical use of Foucault can contribute to the so-called 'project of reflexive modernisation'. Reflexive modernisation refers to an analyses that scrutinises the intentional and unintentional side effects of modernisation, such as environmental pollution, global warming, food shortage and the political and social processes connected to them. 'Actors reflect on and confront not only the self-induced problems of modernity, but also the approaches, structures and systems that reproduce them' (Hendriks and Grin, 2007: 335).

Foucault would surely propose to go deeper into the rabbit hole and state that this reflexive analysis should also be unleashed on the transition management discourses themselves. Each form of knowledge production that has a strong link with politics and society has its dangers, no matter how innocent and how good its intentions may seem. Or, in the ruthless words of Foucault, 'The relationship between rationalization and excesses of political power is evident. And we should not need to wait for bureaucracy or concentration camps to recognise the existence of such relations' (Foucault, 1994a: 328). Harsh words, but what do they imply for the transition manager or researcher? To put it very simply, the road to hell is paved with good intentions. Always continue to critically scrutinise transition management practices, especially the ones you are involved in yourself.

References

Duineveld, M, R. Beunen, K. Van Assche, R. During and R. Van Ark (2009) The relationship between description and prescription in transition research. In: K.J. Poppe, C. Termeer, M. Slingerland (eds.). Transitions towards sustainable agriculture and food chains in peri-urban areas. Wageningen Academic Publishers, Wageningen, the Netherlands, pp. 309-324.

Flyvbjerg, B. (1998a) Habermas and Foucault: thinkers for civil society. British Journal of Sociology 49 (2): 210-33.

Flyvbjerg, B. (1998b) Rationality and Power: democracy in practice. University of Chicago Press, Chicago, IL, USA.

Flyvbjerg, B. (2002) Bringing power to planning research: one researcher's praxis story. Journal of Planning Education and Research 21: 353-66.

Flyvbjerg, B. and T. Richardson (2002) Planning and Foucault: in search of the dark side of planning theory. In: P. Allmendinger and M. Tewdwr-Jones (eds.). Planning Futures: new directions for planning theory. Routledge, London and New York, pp. 44-62.

Foucault, M. (1969) The Archeology of Knowledge. Random House, New York, NY, USA.

Foucault, M. (1973) The Order of Things. Random House, New York, NY, USA.

Foucault, M. (1979) Discipline and Punish: the birth of the prison. Penguin Books, Harmondsworth, UK.

Foucault, M. (1985) The Use of Pleasure: the history of sexuality. (Volume 2). Random House, New York, NY, USA.

Foucault, M. (1986) The Care of the Self: the history of sexuality. (Volume 3). Random House, New York, NY, USA.

Foucault, M. (1994a) Power: essential works of Foucault 1954-1984. (Volume 3). The New Press, New York, NY, USA.

Foucault, M. (1994b) The subject and power. In: J.D. Faubion (ed.). Power: essential works of Foucault 1954-1984. (Volume 3). The New Press, New York, NY, USA, pp. 326-348.

Foucault, M. (1994c) Truth and juridical forms. In: J.D. Faubion (ed.). Power: essential works of Foucault 1954-1984. (Volume 3). The New Press, New York, NY, USA, pp. 1-89.

Foucault, M. (1997) Ethics: subjectivity and truth; the essential works of Michael Foucault, 1954-1984. Allen Lane, London, UK.

Foucault, M. (1998) The Will to Knowledge. The history of sexuality. (Volume 1). Penguin Books, London, UK.

Foucault, M. (2001) Madness and Civilization: a history of insanity in the Age of Reason. Routledge, London, UK.

Foucault, M. (2003) Society Must be Defended. Lectures at the College de France, 1975-1976. Picador, New York, NY, USA.

Foucault, M. (2004) De orde van het spreken. In: W. Gils, H. Hoeks, L. Kate and A. Van Rooden (eds.). Foucault: breekbare vrijheid. Teksten en interviews. Boom, Amsterdam, the Netherlands, pp. 171-179.

Gutting, G. (1989) Michel Foucault's Archaeology of Scientific Reason. Cambridge University Press, Cambridge, UK.

Foucault, M. (1994) The Cambridge Companion to Foucault. Cambridge University Press, Cambridge and New York.

Hacking, I. (1999) Making up people. In: M. Biagioli (ed.). The Science Studies Reader. Routledge, New York, NY, USA, pp. 161-171.

Hall, S. (1997) Representation: cultural representations and signifying practices. Sage, London, UK.

Hendriks, C.M. and J. Grin (2007) Contextualizing reflexive governance: the politics of Dutch transitions to sustainability. Journal of Environmental Policy and Planning 9: 333-50.

Howarth, D. (2000) Discourse. Open University Press, Buckingham, UK.

Kemp R., D. Loorbach and J. Rotmans (2007) Transition management as a model for managing processes of co-evolution towards sustainable development. International Journal of Sustainable Development and World Ecology 14: 78-91.

Koskela, H. (2000) "The gaze without eyes": video-surveillance and the changing nature of urban space. Progress in Human Geography 24: 243-65.

Laclau, E. and C. Mouffe (1985) Hegemony and Socialist Strategy: towards a radical democratic politics. Verso, London, UK.

McHoul, A. and W. Grace (1995) A Foucault Primer: discourse, power and the subject. UCL Press, London, UK.

Scott, J.C. (1998) Seeing Like a State: how certain schemes to improve the human condition have failed. Yale University Press, New Haven, CT, USA.

Van Assche, K. (2004) Signs in Time: an interpretive account of urban planning and design, the people and their histories. Wageningen University, Wageningen, the Netherlands.

Chapter 7

Materiality, nature and technology in agriculture: Ted Benton

Sietze Vellema

7.1 Introduction

Interest in transitions and transition management has gained momentum partly because it offers opportunities for handling persistent problems. A key element in the transition perspective is the change of social-technological regimes. Technological change, and the way society and actors interact with material conditions, are important aspects in the literature on transitions (Rotmans *et al.*, 2005). In particular, placing management of goal-oriented transitions on a roadmap towards sustainable development (Loorbach and Rotmans, 2006) brings to the fore problems in agriculture and food provision, or in energy supply. Both domains relate social action and human behaviour to natural and environmental conditions.

The aim of this chapter is to elaborate on the role in societal change of nature and technology, i.e. the 'materiality' referred to in the chapter title. The chapter centres on the work of Ted Benton of the Department of Sociology, University of Essex. Benton precisely investigates the interaction between social and biological domains. Writing as a sociologist, his background in both philosophy and the natural sciences (especially biology) is quite unusual. Besides his academic work, Benton is also active as field naturalist and ecologist. The review discussion in this chapter suggests that current thinking on transitions may benefit from more explicit theorising about the society-materiality interaction, particularly to understand the limits of intentional processes of change.

7.2 Explaining social change: an argument for materialism

The work of Benton is grounded in the Marxist tradition of historical materialism (Benton, 1977, 1989, 1991, 1993, 1994, 1996a, 1996b, 1997; Benton and Craib, 2001; Benton and Redclift, 1994). Within Marxist scholarly work as discussed by Benton, specific social relations are made dependent on the development of the material forces of production. Accordingly, changes in the forces of production, i.e. new technologies or productive processes, condition how social actors relate. This also entails how human beings relate to natural conditions. Accordingly, social relationships are not entirely socially constructed or an outcome of free will and conscious decision making by human actors. Rather, social relations are a result of so-called 'objective conditions', such as forces of production and nature-imposed conditions of human existence. In historical materialism, the precise configuration between social relations and material conditions represents historical phases, such as capitalism. Spatial variations and historical transformation thus depend on the various ways in which societies interact with nature.

Benton (1989, 1996a) argues that Marx and Engels departed from materialist premises in their philosophy when developing the basic concepts of their economic theory. This suggests a reluctance to recognise nature-imposed limits on human potential and emphasises the transformative powers of technology for human emancipation. In their economic theory, Marx and Engels relied on the classical political economy developed by Ricardo that can be seen as a reaction to the Malthusian argument relating population growth to food security and the available means of subsistence. Ricardo, like Marx and Engels, was reluctant to admit any role for nature-imposed limits and excluded natural scarcity from his analysis of exchangeable value. In this analytical perspective, material conditions do not enter into the costs of production, nor do they impose natural limits on production or capital accumulation. So Benton's reading of the economic theory of Marx suggests that it underrepresents the significance of non-manipulable natural conditions of labour processes and overrepresents the role of human intentional transformative powers vis-à-vis nature.

In transition terms, the Marxist tradition perceives capitalism as the regime able to transcend the conditional and limited character of earlier forms of interaction with nature. The possibility of human emancipation is made dependent on exploiting the potential of the transformative, productive powers of industry. The progressive role of capitalism was located in the development of the forces of production to the point where a transition could take place from capitalism, with its unjust social relations, to a realm of freedom and welfare. The desire to engage with emancipatory processes seemed to contrast with the recognition of material constraints in a way other than that these needed to be brought under control. This led to an almost Utopian view, shared with other theoretical perspectives, on realising human aspiration unrelated to forces or mechanisms embedded in nature that or not or less malleable by purposeful human action.

In response, Benton proposes to return to the original interest in the labour process, wherein human activity, the subject of work (raw materials or natural conditions) and the instruments, or forces, of production are combined. Below, we will see that Benton proposes making a special case for the intentional structure of agricultural labour processes. At this point it suffices to state that Benton argues in favour of a theoretical perspective on social change and transition that recognises the importance of material conditions posing limits on the transformative powers of human actors. In particular, societal development, and in the context of this volume, purpose-driven transitions, remain tied to sources of energy, raw materials and food derived from nature, and also to a range of non-manipulable contextual conditions.

Benton's argues for a 'green historical materialism' as follows:

> [E]ach form of social/economic life has its own specific mode and dynamic of interrelation with its own specific contextual conditions, resource materials, energy source and naturally mediated unintended consequences (form of 'waste', 'pollution', etc.). The ecological problems of any form of social and economic life would have to be theorized as the outcome of this specific structure of natural/social articulation... These conditions and limits have real causal importance in enabling a range of social practices and human purposes which could otherwise not occur, and also in setting boundaries and limits to their sustainability. (Benton, 1989: 77-78)

Consequently, transition or societal change is theorised as a newly articulated combination of specific social practices and specific complexes of natural conditions, resources and biophysical mechanisms. What constitutes a genuine natural limit for one such form of nature/society articulation may not constitute a limit for another. For example, a society that shifts its resource base may effectively transcend what previously were encountered as limits. Hence, in these terms transitions may be about transforming the prevailing pattern of nature/society interaction. Development then is not a unilinear process of quantitative expansion of the forces of production, but rather a qualitatively different way of realising human social possibilities within a context of natural limits and material conditions. Each historical social formation has its own environmental contradictions and social and political cleavages.

7.3 Complementarities between materialism and transition thinking

The work of Benton reveals a strong interest in materiality. The more recent literature on transitions discussed here shares this focus on technology and environmental conditions of societal change. This section describes briefly how the transition literature perceives the social-material interaction. Subsequently, it discusses how Benton locates this discussion in a long-standing dichotomy in science between nature and society, which he considers unproductive for gaining an in-depth understanding of how the interaction works. Benton's epistemological position may complement the scholarly work on transitions because he is more specific about the implications for social science of taking materiality seriously.

Transition management centres on correcting tenacious system errors, which are, for example, embedded in path-dependent technological trajectories. The literature shows a strong interest in understanding this non-linearity and changes in the direction of technological trajectories (Geels and Raven, 2006). Geels and Raven (2006) emphasise the persistency of patterns of technological change and relate this to cognitive rules and expectations guiding technological search and selection. Accordingly, a new direction in technological change also implies a change in the content of rules and expectations. Building on strategic niche management literature, Geels and Raven (2006) argue that new technologies emerge in 'protected spaces' or niches. What happens in these niches or experimental settings impacts evolutionary change through interaction with selection environments and social networks. These protected spaces also provide room for learning and articulation processes with regard to technical design, user preferences or regulatory requirements. According to these authors, the development of generic technological opportunities will materialise in local configurations that work through *bricolage* activities that align the artefact dimension of technologies with, amongst others, tacit skills, available resources and people. This process of alignment can be followed by aggregation activities, including standardisation, codification and model building, which may, eventually, constitute a newly emerging technological trajectory.

This literature on strategic niche management makes technological change an essential element of societal change. The non-linearity of technological change is explained by interactions and alignment between different actors at the local and meso levels. Localised projects provide an important source for grasping the new windows of opportunity. It is in local projects that the multiple components of configurational technologies get to work together and where specific functionalities are discovered, linked to major regime problems such as climate change or emissions.

Geels (2004) widens the scope of the analysis from artefacts and individual organisations to systems. The systemic nature of technological change is related to processes of interaction and cooperation and of competition and selection. Knowledge flows and competences constitute technological systems. He also broadens the scope by considering how technological systems perform social functions; the actual use of technological artefacts becomes a central element, and not only the development of innovative technologies. Human beings are surrounded by technologies and material contexts. These are not merely neutral instruments, but also shape our perceptions, behavioural patterns and activities. Technological artefacts and material conditions form a context for action, and the hardness of technologies and material arrangements may be difficult to change and limits the interpretative flexibility of artefacts. Moreover, rules and institutions, according to Geels (2004), can be embedded in artefacts and material practices.

Consequently, a technology regime, as defined by Rip and Kemp (1998), is the rule set or 'grammar' embedded in a complex of engineering practices, skills and procedures used in production processes, ways of handling relevant artefacts and persons, and ways of defining problems. These practices are embedded in institutions and infrastructures. Accordingly, Geels (2004) proposes using the concept of socio-technical regime in order to be able to include rules and institutions embedded in mutual interdependencies in the analysis and to understand coordination between technological regimes and other, e.g. market, policy or cultural, regimes. This also complicates the task of system builders (Hughes, 1987), who may be seen as transition managers, because they have to handle both the lack of technological malleability and the structural and persistent nature of institutions.

The concept of regimes is important for understanding stability and change. Path dependence in a socio-technical regime is an incentive for incremental innovations in socio-technical systems, leading to or confirming particular paths or trajectories. A question for transition management then is how radical innovations or regime changes emerge, which brings us back to the previous discussion on the activities in niches that deviate from rules and institutions in existing regimes. These may, eventually, reconfigure or replace an existing regime.

The term 'socio-technical landscape', used in the multilevel perspective in the transition literature, is a metaphor for increased structuration of activities through the hardness of both material and social aspects. Material environments, shared cultural beliefs, symbols and values are hard to deviate from (Geels, 2004: 913). However, changes at the landscape level, for example, climate change, may put pressure on regimes, opening opportunities for local experiments or novelties. Also, negative externalities, such as environmental impacts, health risks or safety concerns, and the way they are problematised by, for example, social movements, may lead to pressure on the regime and induce behavioural change among the professionals or practitioners acting within the rules and institutions of the socio-technical regime.

Consequently, transitions are conceptualised as changes from one socio-technical regime to another, wherein both landscape pressures and niche innovations drive change. This highlights the technical, physical and material backdrop that sustains society (Geels and Schot, 2007). Geels and Schot (2007) want to include both the relatively static elements of such a socio-technical landscape, comparable to soil conditions, rivers and mountain ranges, and the dynamic aspects of the external environment, such as rainfall patterns. Van Driel and Schot (2005) distinguish between unchanging or slowly changing factors (climate), long-term changes (industrialisation) and rapid external shocks (wars or price fluctuations). Such landscape developments do not mechanically impact niches and regimes, but need translation by actors to exert influence. This refers to economic and technical actions.

The above seems to align with Benton's more explicit agenda to overcome the persistent dichotomies or categorical oppositions: mind-body, culture-nature, society-biology, meaning-cause and human-animal. He suggests that these types of 'dualist' oppositions constitute obstacles to building viable integrative research programmes creating conceptual room for organic, bodily and environmental aspects and dimensions of human social life. However, he does so without abandoning the achievements of the 'founding figures' (Weber, Marx, Durkheim) of the modern social sciences in defence of the autonomy and specificity of the study of human behaviour and social life vis-à-vis the life science specialisms (Benton, 1991).

Hypothesising a possible critique from Benton of the transition literature, this literature perhaps makes little contribution to theorising about the interface between human social practices and material conditions and consequences. In my reading of the transition literature, the antagonisms in science can still be observed with on the one hand a tendency towards technological determinism and on the other hand a preference for an over-socialised interpretation of change. Here, Benton's epistemological work discussed next may be informative and lead to a more explicit analytical framework that closes the social-material divide and unravels interdependencies between human action and material conditions and natural limits. The possible value of such an endeavour is discussed in the ontological work of Benton on the agricultural labour process and is presented in the final section.

7.4 Contradictions between materialism and transition thinking

Benton's work is a critique of a 'technological determinist' or 'natural reductionist' view, according to which technological changes are represented as exogenous to human society, and of an 'over-socialised' view, which claims that all notions of nature and technology are symbolic constructs of some culture or another (Benton, 1994). The technological determinist or optimist view builds on the idea that there is an autonomous historical tendency for scientific knowledge to grow and accumulate, and that technological innovation is the medium through which this knowledge is applied. This is understood as a means of controlling nature and of making nature serve human purposes. The social, cultural and environmental consequences of this model have received strong criticisms from various perspectives. Benton may ask whether some of the arguments brought forward in the literature on transition management, for example, a preference for environmental managerialism and clean technologies expressed by the Club of Rome, still contain a similar commitment to technological determinism.

Naturalistic determinism conceptualises humans as a species of natural beings. Humans are connected to other elements in the biosphere by a seamless web of interdependencies. This implies that principles underlying ecology can be generalised as a set of norms for human conduct, abstracted from socially produced differentials. Another line of naturalist determinism is found within ecological perspectives emphasising that the human harmony with nature has been destroyed by the development of social hierarchies; the cure contained in this argument is a return to materially more simple, egalitarian and decentralised communities.

Benton's work takes on board the argument that human beings are a species of living organisms, bound together with other species and biophysical conditions in immensely complex webs of interdependence, but he emphasises that the life sciences are insufficient for understanding human interpersonal and social life. Moral agency, social coordination, intentionality, normative regulation and symbolic communication are distinctive features of human social life that require distinct conceptual markers. He (Benton, 1991) argues that the combination of symbolic communication and normative regulation of activity is both a distinctive human set of capacities and, at the same time, a distinctively human need due to the vulnerability to environmental degradation and dislocation. Humans have a unique ability to enhance the carrying capacity of their environments by making use of tools and instruments and by social coordination of activity. History reveals a range of variable 'material cultures' that are culturally mediated and ecologically bounded. Each form of society is characterised by its own specific constellation of limits, affordances and vulnerabilities to unintended ecological consequences. An outcome of this argument is that there may be numerous possible, but qualitatively distinct, directions for future sustainable development within ecological boundary conditions.

A strong interest in the governance and managerial aspects of transition processes may move the discussion away from the above focus on materiality. Such an interest in governance and political processes is also reflected in some of the sociologically informed transition literature (Bos and Grin, 2008; Hendriks and Grin, 2007), which complements a mere emphasis on new equilibriums as responses to system problems. Yet, this literature also has the danger of presenting an over-socialised view of transitions. The debate on future directions of sustainable development then essentially becomes a matter of politics and morals, analytically remote from the reality in the interface between human social practices and their material conditions and consequences.

Over-socialised views of humanity and nature are a consequence of the sharp nature/society dichotomies found in 'classical' sociologies. Accepting the nature/culture dichotomy, according to Benton, either confines social analysis to a domain in which symbolic communication and normative regulation are abstracted from their material supports or effects, or it widens the scope of social analysis to include human embodiment and disposal over material resources. In the latter case, over-socialised views tend towards a reduction of the material world to its symbolic investment within human cultures. Such social analysis 'remains insufficient in so far as it is unable to grasp the ecological and social consequences of *unacknowledged* conditions of social practices in relation to nature, and their unintended or unforeseen consequences' (Benton, 1991: 45, emphasis in original).

The expose above is particularly relevant to recent thinking on purposeful transitions, which is importantly triggered by environmental crises, such as climate change, or the unintended consequences of technological change, such as animal diseases or food scares related to intensive agriculture. Hence, coming to grips, in theoretical terms, with the ways social and material processes interact seems to be essential for managing transitions. However, this endeavour receives less analytical priority in the works on transition management discussed here in comparison with the analysis of (evolutionary) processes and the management and governance thereof.

Benton emphasises the relevance of incorporating the level of inelasticity of natural limits and material conditions into understanding the practices constituting a socio-technical regime. Moreover, he perceives socio-technical regimes as a spatially and historically specific outcome of the interdependencies between human action and environmental conditions. Following Benton's argument, a possible pitfall of transition management literature may be its considering a transition from one socio-technical regime to another to be the result of cultural and social drivers, without acknowledging the biophysical and natural conditions upon which human action, and thus technology selection, is contingent. Benton's discussion proposes that '[h]uman beings cannot successfully adapt to every possible set of environmental conditions, nor are their bio-physical or social environments limitlessly malleable' (Benton, 1991: 24).

Therefore, in the perspective developed by Benton and Redclift (1994), social science needs to find a way to develop explanatory strategies which are not indifferent to the particularities and contingencies of space and time. How can the relationships between humans and the rest of nature be investigated without abstracting these from their physical embodiments? These authors conclude that a full integration is needed of spatial and temporal aspects of social and ecological processes in order for social science to get beyond the inherited nature/society dualism. A subsequent implication is that, due to the nature of environmental processes, social science must also advance beyond the boundaries within which legal and political decision making or authorities operate. Their argument regards it as necessary 'to analyse and explain the ways in which human social structures bind together not only people but also (non-human) animals, physical objects, spatial "envelops", and so on' (Benton and Redclift, 1994: 9). They argue that human beings mediate relations with nature, within specific structural contexts.

This agenda seemingly contrasts with a growing interest in a micro-sociology of subjective life, and associated linguistic and cultural processes, that barely acknowledges the independent reality of 'nature' and the 'environment'. Benton suggests that a middle way, recognising the independent reality of both society and nature, makes it possible to study the interconnection. To avoid sliding into an 'over-socialised' position and to avoid abstractions of society or nature, he investigates the interconnection in terms of specific social practices along with environmental conditions (e.g. physical space and raw material) and media (tools, technologies and bodily activity). This renders natural limits not an external physical reality but a material condition of social practice.

7.5 An application: materiality in agriculture

Benton (1989, 1996b) translates the theoretical discussion described above into an account of the agricultural labour process. The abstract concept of labour processes in historical materialism is the notion of a raw material undergoing a transformation to yield a use value. This transformation is the outcome of a human labour involving the utilisation of raw materials and instruments. The labour process encompasses both intentional human activity and a range of distinct materials, substances and other non-human beings and conditions.

The emphasis in Benton's conceptualisation lies on the purpose of the labour process itself – the intentional structure –, rather than on the material characteristics. The conceptualisation of the labour process in the Marxist tradition represents a broad and general type with a transformative intention structure. Benton's account assumes that agricultural labour processes take place on clear and prepared land, 'using seed or stock animals which already embody past labours of breeding and selection' (Benton, 1989: 67). He contrasts agricultural labour processes with transformative ones (Benton, 1989: 67) because,

human labour is not deployed to bring about an intended transformation in a raw material. It is... primarily deployed to sustain or regulate the environmental conditions under which seed or stock animal grow and develop. There is a transformative moment in these labour processes, but the transformations are brought about by naturally given organic mechanisms, not by application of human labour. (Benton, 1989).

Therefore, Benton concludes that the intentional structure of agricultural labour processes is eco-regulatory. This creates a tension between the forms of calculation by economic agents and economic dynamics in modern agriculture. These dynamics largely resemble a productive-transformative model and the constraints imposed by the intentional structure of an eco-regulatory labour process. Benton argues that this tension is the primary source of ecological problems in modern agriculture.

Benton (1989: 67-68) sets out four distinctive features of eco-regulatory practices as follows:
- Labour is applied to optimise the conditions for transformations in organic processes.
- Labour is aimed at sustaining, regulating and reproducing, rather than transforming.
- The spatial and temporary distribution of labouring activity is shaped by the contextual conditions and the rhythms of organic development processes.
- Nature-given conditions figure both as conditions of the labour process and as subjects of labour.

'These...features draw attention to the dependence of eco-regulatory practices upon characteristics of their contextual conditions' (Benton, 1989: 68). These contextual conditions 'are relatively impervious to intentional manipulation, and in some respects they are absolutely non-manipulable. For example, the...radiant energy from the sun is non-manipulable' (Benton, 1989: 68). This, Benton continues, confines agricultural labour processes to optimising the efficiency of the 'capture' of this energy by photosynthesising plants, or complementing it by artificial energy sources. Also, recent and foreseeable developments in biotechnology have been introduced with a discourse that rendered marginal, according to Benton, the limits, constraints and unintended consequences of their deployment in agricultural systems.

In the social sciences too the transformative capacity of new technologies seems to be exaggerated, suggesting a transition to an industrialised agricultural system. Proponents of such an argument may find it difficult to discover the precise mechanisms embedded in biological processes or organisms that lead to unintended consequences, such as changing resistance to diseases or vulnerability to environmental stresses. A materialist account proposes recognising that organisms are not mere aggregate expressions of contingently related and freely manipulable genetic particles.

Benton's discussion of Marx's generalisation of the intentional structure of labour processes as transformative and value maximising, also questions how transition literature explains the way labour and technology affect material processes. Benton's conceptualisation of the intentional structure of agricultural labour processes emphasises the dependence on non-manipulable conditions and subjects. Labour adapts to these conditions rather than transforms them. This also relates to the social organisation and divisions of labour attached to specific labour processes. This we may compare with what is called a socio-technical regime. Coming back to technology, Benton's argument suggests conceiving contextual conditions separately from the instruments of labour and distinguishing between technologies that enable a transcendence of naturally imposed limits and technologies that enhance adaptability in the face of natural conditions impervious to intentional action. There are causal mechanisms at work which are extrinsic to human intentionality or purpose. Benton concludes that 'once the dependence of productive labour-processes on contextual conditions is explicitly recognized, then the possibility that they may be undermined by their own naturally mediated unintended consequences is open to investigation' (1989: 74).

In Benton's argument, agriculture becomes both a production branch where the primary appropriation of nature is conducted as economic locus as well as a pressure point to where the ever-growing material requirements of other social practices flow. An intentional structure and form of calculation based on value-maximisation and transformative action may be inappropriate to the sustainability of the practices concerned and even act as mechanisms of ecological crisis generation. In this sense, a materialist account helps to theorise real mechanisms producing crises or the persistent problems that transition management aims to address.

7.6 Conclusion and perspectives for social action

This chapter proposes inclusion of materiality in analysing what at first may appear to be processes of social change. The suggestion in Benton's materialist account is, in the case of agriculture, food provision or energy supply, to recognise the dependency of such a regime on a range of non-manipulable contextual conditions and to see agricultural labour processes as specific social forms of interchange with nature. This interchange remains largely under-theorised in the transition literature discussed in this chapter, which may result in a certain blindness to naturally mediated unintended and unforeseen consequences of specific practices, which also impinge upon the persistence or reproduction of contextual or natural conditions.

Benton looks for opportunities to align the 'old' social justice agenda with environmental or green perspectives originating in what can be called the new oppositional social movements concerned with the environmental consequences of economic growth and modern technologies or with animal rights and welfare. He considers it important to go back to the original interest of Marx and Engels in the material conditions of human life and social organisation in order to overcome the tensions existing between ecological politics and the growing interest in sustainable development with the strive for equitable growth of welfare.

One of the implications following from Benton's arguments may be the need for stronger engagement of actions targeting institutional and socio-economic change with the variety of actions for protection of natural environments or conservation of animal and plant life. Since Benton considers social and natural life as inseparable, it may be important to deepen our understanding of how the precise articulation of both strata in specific contexts opens a variety of transition pathways. This then goes beyond blaming industries or farmers for environmental problems resulting from urbanisation and industrialisation, and it also goes beyond technical fixes in, for example, animal husbandry that are supposed to solve complex problems originating in the historically evolved articulation of nature and society.

Discovering ways of marrying environmental protection and social justice brings a new dimension to specific sites of transformative spaces, which, in terms of the transition literature, may be the niches where more systemic changes start. The arguments developed by Benton indicate that change is linked to a historically specific political economy of the interaction between nature and society. Protecting bumblebees (Benton, 2006) then becomes an enormously challenging task involving managing the levers that can redirect developments, and which may also lead to endeavours to change contemporary socio-technical regimes when settling conflicts on materiality and pursuing new forms of cooperation among humans.

References

Benton, T. (1977) Philosophical foundations of the three sociologies. Routledge & Kegan Paul, London, UK.

Benton, T. (1989) Marxism and natural limits: an ecological critique and reconstruction. New Left Review 51-86.

Benton, T. (1991) Biology and social science: why the return of the repressed should be given a (cautious) welcome. Sociology 25: 1-29.

Benton, T. (1993) Natural Relations: ecology, animal rights and social justice. Verso, London, UK.

Benton, T. (1994) Biology and social theory in the environmental debate. In: M. Redclift and T. Benton (eds.). Social Theory and the Global Environment. Routledge, London, UK, pp. 28-50.

Benton, T. (1996a) The Greening of Marxism. Guilford, New York, NY, USA.

Benton, T. (1996b) Marxism and natural limits: an ecological critique and reconstruction. In: T. Benton (ed.). The Greening of Marxism. Guilford, New York, NY, USA, pp. 157-186.

Benton, T. (1997) Beyond left and right? Ecological politics, capitalism and modernity. The Political Quarterly 68: 34.

Benton, T. (2006) Bumblebees: the natural history & identification of the species found in Britain. Collins, London, UK.

Benton, T. and I. Craib (2001) Philosophy of Social Science: the philosophical foundations of social thought. Palgrave, Basingstoke, UK.

Benton, T. and M. Redclift (1994) Introduction. In: M. Redclift and T. Benton (eds.). Social Theory and the Global Environment. Routledge, London, UK.

Bos, B. and J. Grin (2008) "Doing" reflexive modernization in pig husbandry: the hard work of changing the course of a river. Science, Technology and Human Values 33: 480-507.

Geels, F. (2004) From sectoral systems of innovation to socio-technical systems: insights about dynamics and change from sociology and institutional theory. Research Policy 33: 897-920.

Geels, F. and R. Raven (2006) Non-linearity and expectations in niche-development trajectories: ups and downs in Dutch biogas development (1973-2003). Technology Analysis and Strategic Management 18: 375-92.

Geels, F. and J. Schot (2007) Typology of sociotechnical transition pathways. Research Policy 36: 399-417.

Hendriks, C.M. and J. Grin (2007) Contextualizing reflexive governance: the politics of Dutch transitions to sustainability. Journal of Environmental Policy and Planning 9: 333-50.

Hughes, T.P. (1987) The evolution of large technological systems. In: W. Bijker and T. Pinch (eds.). The Social Construction of Technological Systems: new directions in the sociology and history of technology. MIT Press, Cambridge, MA, USA.

Loorbach, D. and J. Rotmans (2006) Managing transitions for sustainable development. In: X. Olshoorn and A. Wieczorek (eds.). Understanding Industrial Transformation: views from different disciplines. Springer, Dordrecht, the Netherlands.

Rip, A. and R. Kemp (1998) Technological change. In: S. Rayner and E.L. Malone (eds.). Human Choice and Climate Change. (Volume 2). Battelle Press, Columbus, OH, USA.

Rotmans, J., D. Loorbach and R. Van der Brugge (2005) Transitiemanagement en duurzame ontwikkeling: co-evolutionaire sturing in het licht van complexiteit. Beleidswetenschap 19: 3-23.

Van Driel, H. and J. Schot (2005) Radical innovations as a multi-level process: introducing floating grain elevators in the port of Rotterdam. Technology and Culture 46: 51-76.

Chapter 8

Sustainable greenhouse horticulture and energy provision: two regional transition processes compared

Jan Buurma and Marc Ruijs

8.1 Introduction

In 2002 the Dutch government designated special areas for the development of greenhouse horticulture. These development areas combined a variety of goals, such as improved economic viability of horticulture, regional economic development and collective generation of energy, as well as orchestrated action for reducing the environmental burden of food provision. The development areas for greenhouse horticulture can thus be seen as an example of a longer term shift towards more sustainable food production. Some of these areas were quite successful and achieved their targets faster than expected. Others were less successful and fell short of expectations.

This chapter focuses on two greenhouse development areas: Agriport A7 in the north-west part of the Netherlands and Bergerden in the east of the Netherlands. Agriport A7 is a modern project location for large-scale greenhouse horticulture and includes an industrial park for agribusiness and logistics. The project is located in the Province of North Holland, next to highway A7, a 30-minute drive from Amsterdam Airport. Agriport A7 is flourishing. The first 500 ha were sold within two years and the second 500 ha were prepared earlier than planned. Bergerden is a government-supported location for the development of medium- and large-scale greenhouse horticulture. The project is located in the Municipality of Lingewaard, between the cities Arnhem and Nijmegen. Bergerden is struggling uphill. The expected influx of growers from the western Netherlands has been disappointing. Ambitions for the second phase were scaled down, and the energy combination went bankrupt.

The challenge of explaining the success or failure of the development areas lies in seeing through the immediate activities and seeking hidden mechanisms in the modes of cooperation and the steering of development. This chapter aims to do just that. It first reconstructs the two development projects. It then applies the selection of theoretical frameworks presented earlier in this volume to provide a 'stepping stone' to insights that may explain the very distinct outcomes of the two horticultural development area cases.

In the next sections the news coverage of Agriport A7 and Bergerden is described in terms of frequencies of articles in national newspapers. This descriptive analysis shows the relative importance of distinct themes in the two contrasting areas. The analysis also reveals which actor groups had the lead in the development of the two areas. The descriptive analysis is followed by a dramaturgic analysis in which the decisive moments and actors for development are identified. Furthermore, the course of development of the two areas is connected with various social theories. The chapter concludes with a reflection on reasons for the success or failure and lessons learned for the future.

8.2 Reconstruction of two regional development projects

8.2.1 Introduction

Agriport A7 and Bergerden are opposites. In Agriport A7 private-sector actors and private-sector drivers (economics: agro-logistics and natural advantages) dominated the news coverage. In Bergerden public-sector actors and public-sector drivers (environment: energy innovation and relocation policy) dominated the coverage. Table 8.1 shows that the coverage of Agriport A7 in national newspapers started some years later (from 2005) than coverage of Bergerden (from 2002). The news coverage of Agriport A7 peaked in the years from 2006 to 2008. The news coverage from Bergerden was highest in the years 2005 and 2006.

Table 8.2 presents the themes discussed in the news articles on Agriport A7 and Bergerden. The main themes are agro-logistics, natural advantages, relocation policy, location facilities, government support, energy innovation, sector promotion and public concerns. The table shows striking contrasts in the news coverage profile of the two areas. The major themes related to Agriport A7 are agro-logistics (27%), natural advantages (22%), relocation policy (15%) and energy innovation (15%). Government support and location facilities are hardly mentioned in the news on Agriport A7. The major themes in the coverage of Bergerden are energy innovation (35%), relocation policy (24%), location facilities (16%) and public concerns (10%). Here, entrepreneurship and natural advantages are hardly mentioned.

Table 8.3 categorises the news coverage in both areas as pertaining to four actor groups: 'policy and governance', 'research and development', 'production and input supply', and 'trade and logistics'. It indicates a dominance of the private sector ('production and input supply' and 'trade and logistics': 48% + 12% = 60%) in the news coverage of Agriport A7. 'Policy and governance' is represented by 33% of the coverage. In Bergerden the public sector dominated

Table 8.1. Numbers of articles on Agriport A7 and Bergerden in national newspapers from 2002-2008.

	2002	2003	2004	2005	2006	2007	2008	Totaal
Agriport A7	0	0	0	4	11	9	9	33
Bergerden	2	8	6	14	24	3	5	62

Table 8.2. Proportion (%) of themes in the news coverage (2002-2008) on Agriport A7 and Bergerden.

Theme	Agriport A7	Bergerden
Agrologistics	26.8	7.4
Natural advantages	22.0	0.7
Relocation policy	14.6	24.3
Location facilities	6.1	15.5
Government support	0.0	6.8
Energy innovation	14.6	35.8
Entrepreneurship	8.6	0.0
Public concerns	7.3	9.5
Total	100.0	100.0

Table 8.3. Proportion (%) of actor groups in the news coverage (2002-2008) on Agriport A7 and Bergerden.

Actor group	Agriport A7	Bergerden
Policy & Governance	33.3	65.6
Research & Development	6.7	8.1
Production & Input supply	48.3	26.3
Trade & Logistics	11.7	0.0
Totaal	100.0	100.0

the news coverage ('policy and governance' and 'research and development': 66% + 8% = 74%). 'Production and input supply' represented 26% of the news coverage. Having in mind the performance of the two development areas (one flourishing and the other struggling uphill), the hypothesis arises that private driving forces offer better prospects for achieving transitions in greenhouse horticulture than public driving forces. However, to confirm this hypothesis a more detailed analysis is needed.

8.2.2 Agriport A7[3]

In 1997 the European Union's Directorate General for Health and Consumers (DG SANCO) published a green paper presenting basic principles of a new food law. The paper was a response to the recent crises surrounding mad-cow disease and dioxins and the public food supply. A practical consequence of the upcoming law for vegetable packing plants was the need to separate incoming and outgoing product flows. The management of the vegetable trading company Hiemstra in North Holland concluded that to comply with the law they would need to centralise their activities and build a new packing plant within the next five to seven years.

In 1999 the Municipality of Wieringermeer published a vision on the future. It had to find an answer to the economic decline of the municipality. One of the options under consideration was the development of a business centre. The management of the vegetable trading company Hiemstra saw this as a good opportunity for synergy between the company and the municipality.

In 2000 Anton Hiemstra, the founding father of the Hiemstra vegetable trading company, bought 50 ha of land next to highway A7 on the northern fringes of the main vegetable production regions of the Province of North Holland. This was the onset of the building of a new packing plant. He got approval from the local and provincial authorities to further develop the plans for the packing plant and an agro-business centre. The only condition set by the authorities was cooperation with other vegetable growers. The location was named Agriport A7. The managing director of Hiemstra had regular business contacts with growers in the South Holland greenhouse district for the procurement of vegetable plants. During these contacts growers had also exchanged views on the future of greenhouse horticulture. In 2003, Hiemstra came in contact with Frank van Kleef. Since 1997 this greenhouse vegetable grower had been seeking a suitable location to build a large-scale greenhouse. Hiemstra and Van Kleef reached an understanding and together developed plans to build greenhouses next to the packing plant and agro-business centre.

In 2004 Hiemstra bought 100 ha of land next to Agriport A7. This signalled the start of development of large-scale greenhouses for (a consortia of) vegetable growers from South Holland. Hiemstra asked for approval from the local authorities to build the greenhouses next to Agriport A7. This request caused some anxiety among local residents and in the Provincial Federation for the Environment.

---

[3] This review is based on information taken from: http://lexisnexis.academic.nl. This is a database where articles from national newspapers, supplemented by interviews with the commercial director of *Agriport A7* and a member of the Provincial Executive were made available.

In July 2005 the Provincial Executive urged Agriport A7 to cooperate with the province-supported greenhouse development area *Het Grootslag* in Wervershoof; this was a sister project to that in the Bergerden area discussed below. The Provincial Executive wanted to avoid competition between the two development areas, because competition, it was feared, might lead to a financially precarious situation for the project *Het Grootslag* and thus for the province. Agriport A7 and the province agreed that Agriport A7 was to be for large-scale greenhouses and *Het Grootslag* for medium-scale greenhouses. Moreover, Agriport A7 became the logistics centre for *Het Grootslag*.

In August 2005 the development company of Agriport A7 presented the environmental impact report for the greenhouse complex and asked for adjustment of the municipal zoning plan. Agriport A7 got approval for the development of the large-scale greenhouse complex. The plan accommodated the concerns expressed by local residents and by the Provincial Federation for the Environment. It included investments in landscape quality and stipulated maximum emissions of artificial light. These elements in the plan satisfied stakeholders not directly involved in horticulture.

In February 2006, Van Kleef announced the establishment of the Energy Combination Wieringermeer (ECW). The existing electricity network in the region was not geared to greenhouse horticulture. Moreover, the power company was reluctant to construct a high-voltage cable. For that reason the growers decided to organise the electricity infrastructure themselves. The Ministry of Economic Affairs gave its permission for this venture. The transformer station was opened in the presence of several politicians and administrators. The entrepreneurs were praised for their initiative.

In November 2006 Hiemstra, the director of Agriport A7, disclosed the company's intention to expand Agriport A7. All land in phase 1 of the project had been sold. For that reason, preparations for phase 2 had to be moved forward. The local council accepted this acceleration.

In March 2007 the provincial government and chamber of commerce selected Agriport A7 as 'Enterprise of North Holland 2006'. The award was given for outstanding achievement in realising 450 ha of greenhouses in a very short period of time. The award implied commercial and political respect for Agriport A7.

In April 2007 the province of North Holland commissioned a study of the rapidly increasing freight transport on the roads serving Agriport A7. The need and options for upgrading the region's road network were investigated. The provincial government's motivation for the study was to support economic development in the region.

In October 2007 the municipality of Wieringermeer approved accelerating phase 2 of Agriport A7. The municipality decided to immediately start the environmental planning for phase 2, two years earlier than originally scheduled. In doing so, the municipality facilitated the growth of Agriport A7 from 500 ha to 1000 ha.

In February 2008 the Provincial Executive announced its ambition to designate special locations ('hotels') for housing seasonal workers from abroad (workers who assisted in the harvest of horticultural crops). The National Farmers Organisation reacted positively, especially in relation to Agriport A7.

In February 2008 the National Farmers Organisation campaigned against construction of a border lake between the former island of Wieringen and the Wieringermeer polder. The farmers wanted to prevent the loss of high-quality arable land for water sports. Referring to Agriport A7, the organisation claimed that agribusiness could offer much more support to the regional economy than recreation. If relocation proved unavoidable, the farmers hoped to receive reasonable compensation.

The timeline of the news coverage from Agriport A7 shows a coincidence of threats to a trading company and a local government. Hiemstra had to cope with the requirements of the EU General Food Law and the municipality of Wieringermeer had to cope with the economic decline of the region. The two parties managed to transform their challenges into a win-win situation. The success story started with a new logistic centre for vegetables and grew into an impressive greenhouse vegetable production centre. The most striking feature of Agriport A7 is the positive attention to the mutual interests of local government, horticultural entrepreneurs, local residents and pressure groups and local farmers. In fact, Agriport A7 is exemplary for successful public-private partnership.

8.2.3 Bergerden[4]

In 1984 the chambers of commerce in the eastern part of the Netherlands published a report on the future of the already existing horticultural complex in *Over-Betuwe Oost*. They advised concentrating greenhouses in the Betuwe region in a new, modern greenhouse horticulture location. The report was welcomed by the province of Gelderland and STOL (a corporation for stimulation of horticulture in Over-Betuwe Liemers). STOL represented a wide range of stakeholders, including local governments, growers and their organisations and horticulture-related entities such as auctions, banks and input suppliers. STOL started to buy land.

In 1990 the City of Nijmegen suddenly presented a plan for developing a new housing quarter, called *Waalsprong*, on the north bank of the Waal River at a location where many greenhouses were currently located. The greenhouse owners reacted with disbelief, fury and obstinacy.

---

[4] This review is based on information taken from: http://lexisnexis.academic.nl. In this database articles in national newspapers, supplemented with inside information from a provincial newspaper and face-to-face interviews with a growers' representative and a local government project leader were made available.

Their future was in shatters. The growers and their families were left in suspense about the future of their businesses.

In 1994 the province of Gelderland together with the municipalities of Huissen and Bemmel presented a plan for the development of a 600 ha, modern greenhouse horticulture location near Bergerden. The area was planned for the relocation of growers from the *Waalsprong* and other locations in the province. A steering group was formed with representatives of the municipalities of Bemmel, Elst, Huissen and Nijmegen. A survey was announced among growers to assess their willingness to relocate. STOL did not want to take part in this process. It considered the survey a waste of time and money because it already had comprehensive information on the growers in the region.

In February 1995 the City of Nijmegen presented the development plan for the *Waalsprong*. The plan included a project for building houses on the edge of a central pond. To realise the plan, the new (publicly funded) greenhouse location *'t Visveld* would have to disappear, just two years after its official opening. The greenhouse growers became hostile and organised a blockade of the Waal Bridge. STOL condemned the waste of public funds.

In May 1995 the steering group organised a news conference to present the (already leaked) results of the survey among the growers. The survey made clear that the growers wanted clarity now. For that reason the steering group wanted to hurry and commissioned the City of Nijmegen to draw up a zoning plan for 90 ha of greenhouse horticulture. The steering group chairperson assumed that greenhouse building would start before the end of the year. Yet already during the news conference a controversy among stakeholders arose on compensation for earlier investments in infrastructure. Two years later the first greenhouse still had yet to be built.

In December 1995 the Municipality of Bemmel presented a draft version of the zoning plan for the Bergerden greenhouse location. Bergerden was envisioned as a modern greenhouse complex, with collective facilities for irrigation water and energy and an attractive landscape. Realisation of the greenhouse park was said would take 10-15 years. The municipality feared the financial risks of developing the location. The province refused to give guarantees. The National Growers Organisation supported the greenhouse location. The plan was approved in early 1999.

In June 1999 local farmers requested the Council of State to call a halt to the zoning plan. According to the farmers the zoning plan shut the door to continuation of their farming operations. They also said that the province had failed to draw up an environmental impact report and that the municipalities were not allowed to require minimum areas for greenhouses. As a result the province set about drafting an environmental impact report, and the Council of State finally (in the course of 2004) disallowed the local farmers' objections.

In 2000 the Ministry of Agriculture concluded an agreement with the National Growers Organisation on the development of new greenhouse locations. The Ministry recognised the need for new greenhouse development areas and provided funds for investments in infrastructure and landscape quality. Bergerden was one of ten locations. In 2002 Bergerden received €10.3 million of STIDUG funds (government subsidy to stimulate sustainable greenhouse horticulture development). The funds were used for location design, infrastructure (water and energy) and landscape quality. The growers paid relatively low prices (€35/m^2) for the greenhouse land. Bergerden was held up as an example for other greenhouse locations in the country.

February 2004 marked the start of the Energy Combination Bergerden (ECB). Director Piet Middelbrink explained that ECB would organise the energy infrastructure in Bergerden and take care of energy generation and distribution. Growers in the greenhouse area were obliged to participate in ECB. In return, they were guaranteed a 10% reduction of their energy costs. Some months later Minister Veerman of Agriculture commented at the opening of Bergerden that inventiveness and persuasiveness would be needed to attract growers from other regions. He was confident, however, that the collective facilities for water and energy would trigger an influx of growers from elsewhere.

In 2005 and 2006 ECB Chairperson Stef Huisman frequently cheered the application of innovative energy systems in Bergerden. He even emphasised the prospect of a 20-30% reduction in energy costs. He furthermore highlighted the use of renewable energy sources (bio-oil and biogas) and novelties like the development of an energy-producing greenhouse. Nevertheless, the influx of growers from western provinces remained disappointing (4 out of 32 growers). Consequently the Ministry of Agriculture was forced to renew the term of operation of the STIDUG funds.

In February 2005 the National Council for the Countryside (*Raad voor het Landelijk Gebied*) advised the government to focus on so-called 'greenports' for horticultural products. These were bundled complexes for transport, shipping and logistics of products in a region. Three greenports were proposed: Westland, Aalsmeer and Venlo. Bergerden was said to offer little prospect as a greenport. The Provincial Executive disagreed with the Council's opinion on Bergerden. It continued to develop plans for further expansion of Bergerden as a greenhouse location and concluded that government support was needed.

In June 2005 the Provincial Federation for the Environment lodged objections to further expansion of Bergerden. The provincial plan to clear the way for 1000 ha of greenhouses in Bergerden was said to threaten the quality of life of the local population. The plan was also said to be incompatible with the planning of *Over-Betuwe Park*. The municipality of Lingewaard pushed the objections off the table, as it saw enough options to maintain landscape quality.

In January 2006 the action group 'Stop the Greenhouses' presented 1,200 signatures to the Royal Commissioner (the chairperson of the Provincial Executive). The action group said that enough space was already available for new greenhouses in Bergerden. Further expansion would harm the quality of life of the population in the neighbouring villages and cities. The Provincial Executive nevertheless decided to clear the way for the 1000 ha expansion of Bergerden in an easterly direction. The growers had advocated expansion in the westerly direction.

In February 2006 the Provincial Federation for the Environment presented an inventory by the Agricultural Economics Research Institute (LEI) confirming the very limited need for additional greenhouses: 75% of the growers had indicated that they had no intention to expand. Politicians in The Hague disagreed with the LEI report. The Provincial Executive saw no reason based on the report to reconsider its earlier decision.

In March 2006 the Provincial Federation for the Environment campaigned in front of the provincial government building. The aim was to convince the members of the Committee for Spatial Planning that expansion would be harmful to the environment and that there was in fact no need for expansion. The Committee supported expansion of Bergerden in the easterly direction. Two weeks later the Provincial Council decided to keep the door open for expansion in 2010, and then in the westerly direction (as advocated by the growers).

In January 2008 creditors of ECB, the collective energy provider, filed a petition for suspension of payments. One month later ECB went bankrupt. ECB had a conflict with the contracting firms on the prices for pipes and cables, which were much higher than estimated. The automation system also turned out to be much more expensive than estimated (€2 million instead of €0.5 million). As a result new growers stopped coming in because of the high costs of the collective infrastructure for energy and water. The 17 growers in Bergerden established a new energy cooperative 'Greenhouse Energy'. They bought the energy infrastructure without the burden of debt. In September 2008 only 110 ha of the 215 ha of land available for greenhouses had been sold.

The timeline of the news coverage of Bergerden shows a long period (1984-2006) of efforts to concentrate the greenhouses in the Betuwe area in a new, modern greenhouse horticulture location. The decision-making process grew lengthy because of bureaucratic procedures and local authorities' fear of financial risks. The provincial decision to put STOL on the sidelines caused a break between the public-sector actors and the private-sector parties involved in horticultural development. The 2000 agreement between the Ministry of Agriculture and the National Growers Organisation was a breakthrough. The €10.3 million of STIDUG funds made investments possible in location design, infrastructure and landscape quality. In 2005 regional and local pressure group campaigns stopped the provincial plans for a further expansion to 1000 ha. The most striking feature of Bergerden is the lack of attention to the mutual interests of local government, horticultural entrepreneurs, pressure groups, and local residents and farmers. The development history of Bergerden presents a clear example of compartmentalisation of the various actor groups involved. Both the government and the growers made many efforts to ensure that Bergerden was attractive to growers from western provinces. In fact, these efforts failed. The showpiece of a collective energy and water supply turned out to be too expensive. The axiom of 'centralisation has advantages' turned out to be false.

8.3 Regional horticultural development areas in terms of transition management

This section presents the case studies in terms of transition, particularly emphasising how dynamics at different levels interfered. Special attention is paid to the historical context of the events, changes in the conditions feeding the processes, the changes achieved in socio-technical regimes, the actors that tried to control the transitions and the moves that turned out to be decisive.

8.3.1 Agriport A7

Concurrent developments: seeds of change

In Agriport A7 a coincidence in circumstances contributed to the success of the business initiative. The story started with an agribusiness entrepreneur confronted with the requirements of the EU General Food Law. Those requirements forced him to identify a new location for a vegetable packing plant. The entrepreneur personally knew a farmer at the location where he wanted to establish the new packing plant. He managed to buy the farm land within days. From that point onwards, the entrepreneur started to include public and private partners in his initiative.

Concurrently, the municipality had established a policy plan for strengthening the municipal economic structure through agribusiness development. So the private sector and the public sector had common interests. Somewhat later, the agribusiness entrepreneur got in touch with a greenhouse entrepreneur who was desperately seeking a large 50-100 ha location. Such a property was relatively easily obtainable next to the new agribusiness centre. Moreover, the provincial government was faced with objections to a new greenhouse location in another part of the province. Again Agriport A7 and local government had a symbiotic relationship.

New arrangements

The agribusiness entrepreneur managed to combine his initiative with the municipal policy plan for agribusiness development. He was prepared to pay consultants, who were tasked to prepare the revision of the zoning plan. The agribusiness centre turned out to be budget neutral for the municipality. Furthermore, the entrepreneur bought the land for the greenhouse entrepreneur without calculating service costs. They operated as business partners. Furthermore, the entrepreneurs tackled the objections of environmental pressure groups before they were dealt with in the municipal council. Getting electricity infrastructure for the greenhouse complex was a problem, because the energy companies did not cooperate. For that reason the energy association of the greenhouse growers realised the infrastructure themselves, thus bypassing the powerful energy companies.

Leading actors controlling the transition process

The entrepreneurs in Agriport A7 managed the transition process themselves. They had an excellent network at several levels (local, provincial and national) and proactively made contact and reached solutions with potential opponents. They bought land before applying for revisions of zoning plans. They helped land owners by offering them new land. They donated funds for nature and landscape development in the region. They did not have a rigid master plan, but always tried to find partners and opportunities for taking further steps.

Conditions for the transition process

These networks at the various levels of government and with business partners in the horticultural sector (conditions at the regime level) provided fertile grounds for localised action and cooperation. The liberalisation of the energy market cleared the way for own investments in energy infrastructure (conditions at the regime level). Economic pressure at the local level stimulated symbiotic relationships between public and private partners (conditions at the landscape level). Creative entrepreneurship and building a critical mass were the basis for everything (conditions at the niche level), rather than having ambitious and complex master plans. In fact, the development process in Agriport A7 was managed as a business development expedition.

In conclusion, the development of the greenhouse location Agriport A7 in Wieringermeer was boosted by conditions and initiatives at the macro level (the socio-economic landscape), the meso level (the socio-technical regime) and the micro level (technical and social novelties). Figure 8.1 positions the conditions and initiatives concerned in the multi-level perspective of transition theory.

The conditions at the macro level are the economic decline in the Municipality of Wieringermeer (in the 1990s) and the development of the new EU legislation on food safety (1997-2004). The conditions at the meso level are the liberalisation of energy markets in the late 1990s and the provincial regional economic stimulation programme (from 2000). The initiatives at the micro level are the very idea for the development of the agribusiness park, the deal with the Provincial Federation for the Environment and the idea for an own power network.

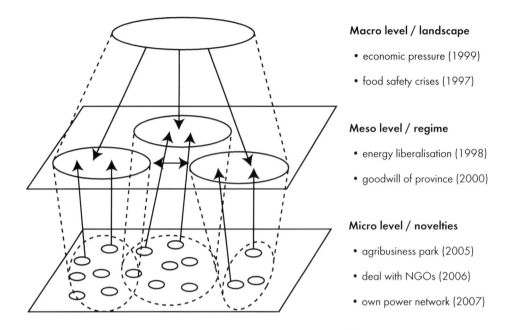

Figure 8.1. Multi-level perspective of the transition process in Agriport A7 (adapted from Geels, 1997: 58).

Macro level / landscape
- economic pressure (1999)
- food safety crises (1997)

Meso level / regime
- energy liberalisation (1998)
- goodwill of province (2000)

Micro level / novelties
- agribusiness park (2005)
- deal with NGOs (2006)
- own power network (2007)

The dates show that the initiatives resulted from the conditions at the macro and the meso level. The novelties show that the entrepreneurs involved were thinking more broadly than their own business (vegetable trading and logistics and vegetable production). They were prepared to organise an agribusiness park, to make a deal on landscape and to organise their own power network.

8.3.2 Bergerden

Concurrent developments: seeds of change

The idea to develop a new, modern greenhouse location in the Betuwe area was launched in a 1984 report by the chambers of commerce. Four years later (in 1988) two members of the Provincial Executive took the initiative for Bergerden. They founded STOL (a corporation for the stimulation of horticulture in Over-Betuwe Liemers). The task of the corporation was to acquire land for the relocation of greenhouses from other parts of the province. STOL was chaired by a greenhouse entrepreneur and received grants for the acquisition of land from both provincial and local governments and sector organisations.

In the 1990s, two concurrent developments actually formed the basis for the greenhouse project location in Bergerden. First, there was a growing social aversion to solitary greenhouses in the agricultural landscape in the province of Gelderland. Second, the expansion plans of the city of Nijmegen (*Waalsprong*) necessitated relocation of an existing greenhouse centre. These developments took place in the 1990s. In addition, the project location Bergerden was meant to provide space for the relocation of greenhouses from elsewhere in the province. In 2000 Bergerden was recognised as an agricultural development area for greenhouse horticulture, giving the project access to stimulation funds from the federal government.

New arrangements

In 1994 (when the city of Nijmegen launched the *Waalsprong* housing development) the activities of STOL were taken over by the Steering Group on Greenhouse Horticulture (with representatives of the municipalities of Nijmegen, Huissen, Elst and Bemmel). The diversity of interests of the four municipalities and fear of the financial risks of the greenhouse location complicated decision making. Due to the recognition of Bergerden as an agricultural development area, the project received access to stimulation funds (STIDUG) from the national government. Simultaneously, requirements were introduced for environmental protection, landscape quality and infrastructure (water and energy). Owing to these requirements and objections from local farmers it took several years (2000-2004) to get approval for the zoning plan. In 2005-2006 the Provincial Federation for the Environment crossed the province in its ambitions for further expansion of Bergerden.

Besides the arrangement concerning planning, a new form of joint action was planned in the establishment of Energy Combination Bergerden. This was an effort to make the project location more attractive to growers. At the end of the day, the collective facilities turned out to be more expensive than estimated. They then became a stumbling block for the sale of building plots for greenhouses.

Leading actors controlling the transition process

In 1994 the steering group took over the tasks of STOL, and STOL was largely marginalised. Sidelining STOL induced negative reactions from the growers, their organisations and horticulture-related entities. STOL had considered it a waste of time and money to recollect the comprehensive information from growers which STOL already had available. The atmosphere further deteriorated when the city of Nijmegen claimed the brand-new greenhouse location *'t Visveld* as land for development of housing. The province at first failed to draft an environmental impact report, further confusing the process.

Finally Bergerden was saved by its recognition as an agricultural development area and the allocation of €10.3 million of STIDUG funds in 2002. In 2006 campaigns of the Provincial Federation for the Environment forced the Provincial Executive to give up its ambitions for further expansion of Bergerden in an easterly direction. This synopsis shows that in fact, several actors tried to control the transition process, but nobody seems to have had comprehensive control.

Conditions for the transition process

The involvement of municipalities with diverse interests was counterproductive to the quality and pace of decision making. Revision of the zoning plan before acquiring the land provided land owners comfortable negotiation positions. It might also have slowed the pace of the development process. A land ownership bank can be a useful instrument to facilitate the process of land acquisition. Interaction between local government and the horticultural sector is necessary to adjust the design of a project location to the needs of the entrepreneurs. This implies that a greenhouse development project should be managed differently than a housing development project. Bergerden exhibited the features of a housing development project. Also, management of the collective facilities for energy and water complicated the development of the area. Substantial investments in pipes and cables for constructing the hardware of the collective facilities became a stumbling block for the influx of growers. New growers made their own calculations of the potential benefits of centralised service provision, and apparently decided that this was not an attractive offer for moving their operations.

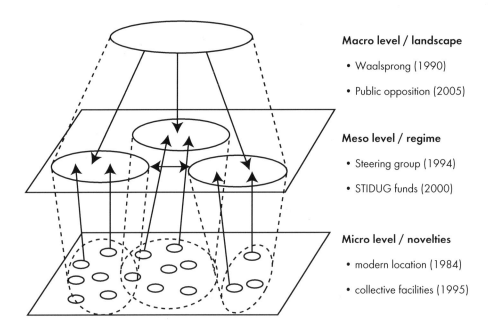

Macro level / landscape

- Waalsprong (1990)
- Public opposition (2005)

Meso level / regime

- Steering group (1994)
- STIDUG funds (2000)

Micro level / novelties

- modern location (1984)
- collective facilities (1995)

Figure 8.2. Multi-level perspective of the transition process in Bergerden (adapted from Geels, 1997: 58).

In conclusion, the development of the greenhouse location Bergerden between Arnhem and Nijmegen was influenced by conditions and initiatives at both the macro level (the socio-economic landscape) and the meso level (the socio-technical regime). Figure 8.2 shows the conditions and initiatives concerned positioned in the multi-level perspective of transition theory.

The idea for Bergerden was born at the micro level. As early as 1984, the chambers of commerce promoted the development of a new, modern greenhouse horticulture location in the region. In 1995 the local governments presented a plan for a modern greenhouse location, with collective facilities for irrigation water and energy and situated in an attractive landscape. In 1990 the plans for the *Waalsprong* placed the ideas for Bergerden in a new context. A place to relocate many greenhouses became urgent. This urgency resulted in the establishment of a steering committee, which put STOL on the sidelines. The development of the Bergerden location gained momentum (in 2000) through stimulation funds from the Ministry of Agriculture. Five years later (in 2005) the plans for a further expansion of Bergerden lost momentum due to public opposition and lack of interested growers from the western provinces.

8.4 Theoretical reflection

This section interprets the patterns discovered in the sequences of events in both areas for the development of greenhouse horticulture from the perspectives of a selection of social theories. Key elements of the different theories are mentioned and practical findings are indicated in the historical reviews.

8.4.1 New institutional economics: North and Aoki

The starting point for new institutional economics is the tactical and strategic behaviour of agents, who have certain beliefs about the world. The leading agents in the development history of Agriport A7 were the vegetable trading company Hiemstra and greenhouse vegetable grower Frank van Kleef. Judging from their actions, that is to say, their developing a business park and constructing a power infrastructure, we can conclude that they considered the development of Agriport A7 to be a business development expedition. Their strategy was to identify favourable conditions (logistics, climate, large scale), create win-win solutions and solve problems before they became troublesome. In their relations with the local and provincial government they reached a new equilibrium in which Agriport A7 had the lead in developing the zoning plan and drafting an environmental impact report. Part of the compromise was that Agriport A7 also accepted partial responsibility for the development of the provincial government-supported greenhouse development area. In return, the provincial government took responsibility for upgrading the road network in the region and for organising housing for seasonal workers from abroad. Apparently, the institutional arrangements in the area for greenhouse horticulture enabled a situation wherein both public and private partners took their responsibilities for the regional economy.

In the case of Bergerden, the leading agents were the municipalities and the provincial government. From their actions we can conclude that they considered the development of the new greenhouse location to be a special version of a housing development project. As a result, their strategy was to take the lead, design a zoning plan, acquire land (sold willingly or unwillingly by the owners), construct infrastructure and sell the lots to the new greenhouse growers. To get support from various stakeholders they promised that a new modern greenhouse location would fit into an attractive landscape. The rules and symbols applied in housing development projects were also applied in Bergerden: e.g. establishment of a steering group, settlement of objections of local residents and land owners, and construction of collective facilities for water and energy. Bergerden can be typified as an example of a government-led development process for a new modern greenhouse location. The growers themselves had a minor role in the process. In contrast to Agriport A7, the institutional arrangements in Bergerden seem less able to accommodate different interests and to work towards a new equilibrium reflected in the use of land.

8.4.2 Social systems: Luhmann

Social systems, according to Luhmann's theory, are self-reproducing networks of communication. Important concepts in this theory are autopoiesis, code of communication, lack of observation and path dependencies. Some of these concepts are present in the development history of Agriport A7. Figure 8.3 represents the social systems observed in Agriport A7.

Figure 8.3 distinguishes three social systems in Agriport A7: (1) entrepreneurs, (2) government and (3) local population, each using their own vocabulary. The entrepreneurs

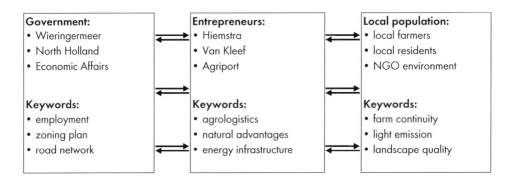

Figure 8.3. Social systems observed in Agriport A7.

are positioned in the centre of the figure, because of their close interactions with both the government (at various levels) and the local population (various groups). Double arrows are drawn between the social systems because of the open atmosphere found in Agriport A7. The codes of communication are quite different between the distinct systems. The entrepreneurs talk about topics like agro-logistics, natural advantages and energy infrastructure. Terms frequently used within the government are employment, zoning plans and road networks. The local population is concerned with farm continuity, light emissions and landscape quality. From a Luhmannian perspective the observational capacity within the distinct social systems appears to have been translated in such a way that it accommodated a joint effort, which gives the impression that the social systems really understood one another.

Figure 8.4 distinguishes three social systems in Bergerden: (1) horticulture, (2) government and (3) local population. The government is positioned in the centre of the figure, because of the opposition of the horticulture sector (1990-1995) on the one hand and that of the local population (1999-2006) on the other hand. Single arrows are drawn between the social systems, reflecting the tense situation found in Bergerden. The codes of communication of the three systems are quite different. The horticulture parties were talking about waste of money, compensation for earlier investments and the need for new locations. The government communicated how a new and modern greenhouse location would fit into an attractive landscape and be well equipped with collective facilities. The local population was concerned with the continuity of their farms, their quality of life and the possible negative impact of horticulture for the environment. Lack of observation capacity was obvious in the government ranks during the debate on the need for further expansion of Bergerden (2005-2006). In 2005 the municipality disregarded the objections of the Provincial Federation for the Environment. In 2006 the Provincial Executive saw no reason to reconsider the decision to clear the way for a 1000 ha expansion of Bergerden, even after receiving a report indicating that 75% of growers had no wish to expand. On the other hand, the Provincial Council translated the signals received from the research institute to alter its original position. A more in-depth analysis following a Luhmannian perspective could further reveal the patterns

Figure 8.4. Social systems observed in Bergerden.

of communication within the boundaries of social systems in order to explain the different outcomes in terms of joint direction in both development areas.

8.4.3 Strong and weak ties: Granovetter

Granovetter's theorising on strong ties and weak ties as conditions for information exchange between actor groups and the diffusion of innovations may help to explain the differences between the areas. There are strong ties among the identified actor groups. However, the observed weak ties, which play a crucial role in social learning, may explain why the development of Agriport A7 advanced quite rapidly. In Agriport A7 there were frequent contacts between institutionally remote actor groups. Hiemstra had frequent contacts with representatives of the municipality and the province. Van Kleef and his colleagues had frequent contacts with the Ministry of Economic Affairs, with the neighbours of Agriport A7 and with the Provincial Federation for the Environment. Although these ties can be characterised as weak, using Granovetter's conceptualisation, they also offered opportunities to borrow from other players and eventually to set in motion a process of social learning supportive to the overall endeavour.

The theory of Granovetter may also explain why the development of Bergerden was struggling uphill. In 1994 the government (province and municipalities) sidelined STOL. As a result, STOL reacted negatively to government plans. The situation improved in 1995 when the National Growers Organisation responded favourably to the plans for Bergerden. This attempt at conciliation led to a breakthrough in the development of Bergerden. The Provincial Federation for the Environment crossed the *no man's land* between government and the local population. It approached the province to get the quality of life of the local residents put on the policymakers' agenda. It also tried to convince the policymakers of the very limited need for additional greenhouses. The Provincial Council finally picked up the arguments and changed the provincial expansion policy. Perhaps a deliberate effort to encourage weak ties between a variety of actors might have removed the obstacles for further expansion and joint action in Bergerden.

8.4.4 Doubt management: Weick

Weick explains that lack of sound doubt obstructs progress of thought. A relentless fixation on one's own convictions prevents people from finding solutions to organisational problems. For that reason organisations need loosely coupled elements, i.e. a flexible learning ability. Sound doubt was found at several places and points in the development history of Agriport A7 and at very few places and instances in Bergerden.

An obvious example of a loosely coupled element in Agriport A7 was the deal with the Provincial Federation for the Environment on investments in landscape. The establishment of an own energy infrastructure was another example of an unconventional activity for greenhouse growers. Furthermore, the deal with local residents on maximum emissions of artificial light can be mentioned as a loosely coupled element. The municipality and the province also deviated from conventions. They delegated the preparation of an environmental impact report to Agriport A7. They honoured Agriport A7 for its achievements with the stamp 'Enterprise of North Holland 2006', which implied commercial and political esteem for Agriport A7.

In Bergerden the local and provincial governments had a strong fixation on the conviction that a greenhouse development location was commensurable with a housing development project. For that reason they sidelined STOL and took the development of the greenhouse location into their own hands. They presented Bergerden as a modern greenhouse development location, with collective facilities for irrigation water and energy, perfectly situated in an attractive landscape. In 2005 and 2006 the local and provincial governments were strongly fixated on the further expansion of Bergerden. They refused to accept the objections of the Provincial Federation for the Environment and saw no reason in evidence presented by research to reconsider earlier decisions. Despite the fixation on the housing project approach and on further expansion, there was evidence of sound doubt. When local farmers lodged their objections to the Council of State, the provincial government rapidly drafted an environmental impact report. After the campaigns of the Provincial Federation for the Environment, the Provincial Council postponed the expansion to some future point in time.

8.4.5 Power and discipline: Foucault

Foucault's conceptualisation of discourse detects four mechanisms: disciplining, subjection, exclusion and evasion. These can be traced in the sequence of events observed in Agriport A7 and Bergerden.

The mechanism of disciplining was applied by the Provincial Executive of North Holland. It urged Agriport A7 to start cooperation with *Het Grootslag*. Earlier, local and provincial authorities had approved the new packing plant at Agriport A7 under the condition of cooperation with other growers. The farmers in Wieringermeer tried to discipline the province by asking for reasonable compensation for their high-quality land in case of construction of the border lake. The mechanism of subjection was applied in a positive way in establishment of the group 'Neighbours of Agriport A7'. Use of the word 'neighbour' here suggests equality and complementarities.

Disciplining was applied in Bergerden by public and private stakeholders who demanded adequate compensation for earlier investments. An element of disciplining was also included in the establishment of the steering group for Bergerden. This meant that STOL no longer had the lead in the development of the new greenhouse location. The mechanism of exclusion was applied in Bergerden with the characterisation 'a modern location in an attractive landscape and with collective facilities.' This implied that anything else was outside the scope of the project. This incantation was no longer accepted as of 2005, when the Provincial Federation for the Environment raised the issue of quality of life of the local residents. The mechanism of evasion was used by the Municipality of Lingewaard, when it stated that there were enough options for maintaining landscape quality.

8.4.6 Materiality: Benton

The work of Benton centres on the interaction between social and biological domains. The idea is that intentional processes of change can be hampered by limits to transformation in the natural realm or by non-malleable biological mechanisms. The suggestion is that social organisation partly results from its embedding in different strata. In the case of Agriport A7, food crises (mad-cow disease and dioxins) and the public response to them (the new EU General Food Law), may explain the origin of surprising alliances and forms of concerted action. In addition, the specific climatic conditions in the Wieringermeer, i.e. more sun radiation and wind, created favourable conditions for a joint endeavour in energy provision. In Bergerden, natural conditions were less attractive, and insufficient to entice the majority of growers to embark on a strategy of expansion. The basis for joint action and the regional horticultural cluster were therefore eroded.

8.5 Conclusions

This chapter began by examining the different dynamics and outcomes of two horticultural development areas. Though they were organised under a similar policy label, the main actors and drivers in the two locations were quite different. In Agriport A7, private-sector actors ('production and input supply' + 'trade and logistics') and private-sector drivers (climatic conditions and agro-logistics) played a dominant role in brokering relationships and establishing social settlements with public authorities. The partners managed to find win-win solutions, they understood and wanted to accommodate one another. In retrospect, Agriport A7 shows several characteristics of continuous change, making use of opportunities and exceptions and updating social processes. In Bergerden, however, public-sector actors ('policy and governance') and public-sector drivers (energy innovation and relocation policy) were dominant, and their behavioural patterns and institutionalised rules unintentionally eroded the foundation for developing the horticultural area. Bergerden is a good example of compartmentalisation: STOL (with both public and private representatives) was sidelined and the different social networks had a strong fixation on their own conventions. Bergerden shows several characteristics of episodic change, based on plans and intentions and also resulting in defensive routines.

This chapter made a first attempt to capture these differences in theoretical terms, which further hypothesised conditions enabling or constraining joint and orchestrated action in horticultural development areas. More theory-informed work is needed to deepen our understanding of the observed paradox in managing these specific transition processes: the episodic change in Bergerden took much more time than the continuous change in Agriport A7, despite a similar intention framed at the national policy level. This suggests that to understand processes of change, it is insufficient to concentrate on recipes and technical or managerial fixes in a bounded arena. Setting in motion a transformation of the agricultural landscape requires a more substantive effort to modify rules and institutions in geographically specific regimes, because change at the micro level or in strategic niches does not automatically translate into a change of the regime.

References

Geels, F.W. (1997) Met de blik vooruit: Op weg naar socio-technische scenario's. Rapport voor Mumford-projecten. Universiteit Twente, Enschede, the Netherlands.

Website

LexisNexis: http://lexisnexis.academic.nl.

Chapter 9

The reconstruction of livestock farming in the Netherlands

Carolien de Lauwere and Sietze Vellema

9.1 Introduction

The outbreak of classic swine fever in the Netherlands in 1997 started a large-scale reconstruction process. This rather top-down land use planning exercise was intended to solve problems and to manage risks in food provision and livestock farming. A key feature of the process was its drastic interference in the spatial and environmental planning of livestock farming. The original aim was to radically restructure pig farming to prevent new outbreaks of livestock plagues in the future. In the course of time, however, the aim of the process was broadened, and more emphasis was put on an integral and systematic approach to the complex problems of spatial planning, environment, nature preservation, landscape, water and economics (VROM, 2003). This led to the Law on Reconstruction of Concentration Areas in 2002, which was designed to improve the quality of the environment (nature, landscape, water, and air quality) and the socio-economic vitality of rural areas. The Dutch national government, county councils and rural municipalities and non-governmental organisations (NGOs) participated in the process. The provincial governments were responsible for composing reconstruction plans with the rural municipalities and social sectors. These plans translated the national government's starting points into actual measures and procedures tailored to the conditions in that particular area (Boonstra *et al.*, 2007).

It took a lot of time before the reconstruction law came into being. Developments in related policy documents about odour and ammonia played an important role in this. Moreover, the Dutch Lower Chamber added a comprehensive zoning plan to the discussions on the law at the provincial government's request. This plan introduced a zoning of rural areas into three types of areas: agricultural development areas, intertwining areas and expansion areas. In agricultural development areas the focus was to be on agricultural activities. In expansion areas emphasis was on residential areas, recreation and nature preservation. In the intertwining areas, there was some room for manoeuvre for farming activities, but these were to be interwoven with residential areas, recreation and nature preservation (Boonstra *et al.*, 2007). Farmers living in intertwining areas or expansion areas who wanted to enlarge their farm operations had little possibility for doing so unless they were prepared to move to an agricultural development area. This, however, appeared to be a daunting endeavour.

This chapter first describes how this general reconstruction process materialised in the Dutch towns of Markelo and Grubbevorst. It then further qualifies the process in both locations in terms of transition, with a special interest in how the regime, e.g. the rules and institutions directing behaviour, was affected or not. This is particularly interesting because of the wide range of protest and unrest that accompanied the change process. Finally, the chapter uses the theoretical inputs presented in the first part of this volume to deepen understanding of the enabling and constraining mechanisms visible in the change processes in both localities.

9.2 The time scale of the reconstruction process

The reconstruction process started in 1997. The Law on Reconstruction of Concentration Areas finally came into being in 2002. But with that milestone, the reconstruction process was not over yet. Actually, it had just started. Boonstra *et al.* (2007) provide an extended overview of the reconstruction process from 1997 to 2005, when realisation of the reconstruction plan really got under way (although pilot projects had been carried out before that time). The reconstruction process is set to be finished in 2016. Before then, however, many obstacles have to be moved out of the way.

Using the Internet tool 'LexisNexis Academic' an overview was made of all major and minor articles in national and local newspapers and journals about the reconstruction process from 1997 until now. The Dutch words *reconstructie* (reconstruction) and *landbouwontwikkelingsgebied* (agricultural development area) were used as keywords. This produced 375 hits from September 1997 until October 2008. The tone of the articles was then analysed and categorised as positive, negative or neutral. Strikingly, the number of articles about reconstruction and agricultural development areas particularly increased from 2003 until 2008 (Figure 9.1). In the early years, when the reconstruction law was coming into being, there was a comparable number of articles with a neutral or negative tone. From 2006, however, the number of articles with a negative tone increased. In that year, the word 'mega-farm' or other related (Dutch) words (e.g. 'pig flat', 'pig invasion', 'pig cluster' and 'pig factory') also showed up in the articles for the first time. Next to this, resistance from local farmers and residents who did not want newcomers in their area played a role (see Table 9.1 and Annex 1 for illustrative headlines).

9.2.1 Markelo and Grubbenvorst: two reconstruction processes

The current study looked in more detail at the reconstruction process in the agricultural development areas surrounding two Dutch towns: Markelo and Grubbenvorst. In Markelo, an organisation called 'family farm plus' (*gezinsbedrijf plus*) bought 40 ha of land to establish a cluster of six farms for about 20,000 pigs in total and a commonly owned company – actually a seventh 'farm' – for manure processing. The 'family farm plus' project was an initiative of LTO-North, a Dutch interest group for farmers, and the feed company 'ForFarmers'. Interested farmers could join the initiative. In Grubbenvorst, there were plans to build two

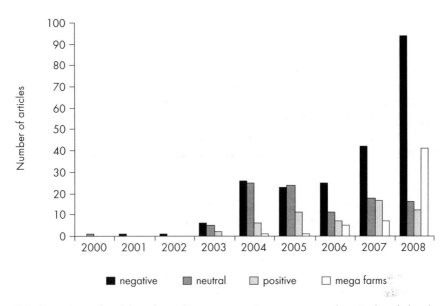

Figure 9.1. Overview of articles about the reconstruction process and agricultural development areas in Dutch journals and newspapers between January 2000 and October 2008.

Table 9.1. Negative headlines in Dutch journals and newspapers about the reconstruction process in Markelo (from July 2006 to October 2008) and Grubbenvorst (from July 2007 to October 2008).

Markelo - family farm plus
    Neighbourhood finds pig cluster much too big (local newspaper, July 2006)
    Suspicion of pig cluster Markelo (national newspaper, July 2006)
    Pigs or peace seekers: coming of six pig farms causes commotion in Markelo (national
      newspaper, December 2006)
    Resistance against big pig farms (agricultural newspaper, March 2007)
    Aggravation of 10,000 screaming pigs; inhabitants of Overijssel use petition to try to keep
      mega-farms out of the landscape (national newspaper, March 2008)
Grubbenvorst - the new mixed farm
    Grubbenvorst awaits puffs and allergies (provincial newspaper, September 2007)
    Bacteria mega-farm threatens horticulture (horticultural journal, January 2008)
    Mega-farm neither mega-bad nor mega-good; four studies not unanimous (provincial
      newspaper, February 2008)
    Farmers win a battle, but fight goes on (provincial newspaper, February 2008)
    Mega-farm: locally more stench and fine dust (provincial newspaper, May 2008)
    Signatures against mega-farms (provincial newspaper, September 2008)

(http://lexisnexis.academic.nl)

mega-farms with 1.3 million head of poultry and 35,000 pigs in a project called the 'new mixed farm' (*nieuw gemengd bedrijf*). The poultry farm implied the inclusion of a slaughter house as well. Manure and slaughter waste would be processed in a common installation for bioenergy, which in turn would produce warmth for a nearby cluster of greenhouses. Initiators of the 'new mixed farm' concept were four innovative agricultural entrepreneurs collaborating with research institutes.

Similar patterns can be observed with regard to the implementation of the plans in Markelo and Grubbenvorst:

- The plans raised many questions among local policy-makers, citizens, farmers and other area residents.
- The plans were approved by the local municipalities with only a very small majority (just three votes in Markelo and one vote in Grubbenvorst).
- Local farmers and residents protested against the plans.
- Independent research was carried out to ease the worries of the protesters.
- The research reduced concerns among local policy-makers but did not help to ease the objections of (most) local farmers and other nearby residents.
- The farmers who wanted to participate in the 'family farm plus' or 'new mixed farm' initiatives received permission to start permit procedures, despite the continuing objections.

In both cases, residents collected signatures and submitted a petition to get their objections against the 'mega-farms' on the agenda of the Dutch Lower Chamber. As of the end of the research period (October 2008), the issue had not yet been resolved.

The survey of Dutch journals and newspapers revealed a lot of negative publicity about the reconstruction plans in Markelo and Grubbenvorst. Between July 2006, when the first article about Markelo appeared in a local newspaper, and October 2008, 28 articles about the Markelo process were written in local and national newspapers and journals, of which 20 had a negative tone, six were neutral and two had a positive tone. The Dutch words *landbouwontwikkelingsgebied* (agricultural development area) and *Markelo* were used as keywords. Between July 2007, when the first article about the reconstruction process in Grubbenvorst appeared in a local newspaper, and October 2008, 80 articles related to the process in Grubbenvorst were written, of which 50 had a negative tone, 20 had a neutral tone and 10 had a positive tone. The Dutch words *nieuw gemengd bedrijf* (new mixed farm) and *Grubbenvorst* were used as keywords.

The examples of negative headlines presented in Table 9.1 suggest that the reconstruction process still has a long way to go until 2016.

9.3 The reconstruction process in terms of transition

The reconstruction process has several features of transitions. It is a long-lasting process, with a time span of almost 20 years. The final goal might be 'more or less' clear, but the way to reach the goal is far from clear and has undergone several changes over the years (Rotmans, 2003). Considerable effort was put into getting public support, which also is a part of a transition process (Grin and Van de Graaf, 1996). The time scales of the reconstruction process itself and of the reconstruction cases in Markelo and Grubbenvorst, however, show that public resistance grew through the years.

9.3.1 Interfering developments

Rotmans *et al.* (2001) define transitions as long-term processes of change during which a society or a subsystem of society is fundamentally altered as a result of interfering developments which strengthen one another in the areas of economics, culture, technology, institutions and nature and environment. In these terms, the reconstruction can be interpreted as a transition, especially in view of the context when the reconstruction came into being. At that point, with the outbreak of swine fever, nature had struck hard in the modern, often highly specialised pig farming sector with its high standards of technology that had enabled pig farmers so far to keep their pigs healthy and keep diseases at bay. This had enormous economic and social impact. All livestock (related) transports were banned and exports were prohibited. Pig stalls became full to the bursting point, and the public was awakened to the problem by intensive media attention and horrible images on television.

The time was ripe for a change, and the reconstruction law seemed to have the answer. From the beginning substantial effort was put into getting public support from local authorities and NGOs with regard to environment, nature and landscape. The reconstruction process went well until the actual reconstruction plans became more concrete. From that point on, protests started. A reason might be that at first the plans were abstract and confined largely to paperwork; once the law came into being it became clear what the reconstruction plans really meant for the involved municipalities and their inhabitants. An agricultural newspaper, for example, headlined about Grubbenvorst: *New Mixed Farm: Interesting Project But Not Here* (September, 2008, http://lexisnexis.academic.nl). In this article, local stakeholders in Grubbenvorst did not appear to be against the concept of the new mixed farm, but they did not want it in their area because a lot of beautiful scenery there had already disappeared for the sake of large-scale construction projects. An October 2007 headline in a national newspaper read, *Fear for 35,000 Pigs in the Backyard* (http://lexisnexis.academic.nl).

Another reason for the increasing protests against the reconstruction process might be that farmers and residents were not adequately involved in the process. An illustration of this is found in a local newspaper article about Markelo headlined, *Everyone Loses in Pig Cluster* (September, 2007, http://lexisnexis.academic.nl). The excerpt below is translated from Dutch:

The pig cluster in Markelo will come. The local council has decided this. A decision without support. ...Should we now cheer? Well, no. This decision only has losers because the residents were not heard. The decision is not supported by the community in Markelo. The people who want this development say that it cannot be stopped and that it is important for the pig sector. But the pig sector is unpleasantly silent. There are only two prospective buyers and they are 'pig cowboys' from Brabant.... People who love nature lose out too. The character of the area will be enormously affected.... There are two pig farms in the municipality that have to be relocated. Nobody is against that. And then wait for further developments. In the near future enough pig farmers will stop farming. Pig farmers who want to move can take over these farms without major disruptions of nature. If that happens the losses could be limited. (http://lexisnexis.academic.nl)

9.3.2 Farm level or micro level

At the micro level, the reconstruction process offered opportunities for expansion-oriented farmers for whom enlargement of their farms was restricted or more or less impossible because they were located in intertwining or expansion areas. Moving to a agricultural development area offered these farmers opportunities for both growth and innovation. Another opportunity concerned environmental advantages and advantages for nature and landscape because farmers would move from areas classified as 'vulnerable' in terms of environment, nature and/or landscape to the agricultural development areas. A problem underestimated so far, however, is that local farmers and other people living in the designated development areas are not always happy with the 'newcomers'. Annex 1 presents more detail on the stories behind headlines such as these: *Farmers Angry About Mega-Farm, No New Farms in Agricultural development area After All, Farmers in Reconstruction Areas Increasingly Diametrically Opposed to Each Other* and *Own Pigs First* (http://lexisnexis.academic.nl).

9.3.3 Regime

The reconstruction plans that had been developed by county councils, municipalities, NGOs and social sectors met a lot of resistance. A possible reason for this opposition is that developments at the micro, meso and macro level did not reinforce one another. This has been suggested as a necessary condition for transition processes (Geels, 2002). First of all, the public did not want mega-farms, and from 2006 the reconstruction process and agricultural development areas seemed to be increasingly associated with mega-farms. In Grubbenvorst, family doctors were concerned about the health of local residents. A provincial newspaper headlined, *Limburgian Doctors See Mega-Farms as a Danger to Humans – the Farmers' Lobby Triumphs* (December, 2007, http://lexisnexis.academic.nl; Annex 1). Furthermore, animal welfare was an important issue, and not all stakeholders involved were convinced that the mega-farms would contribute to improving animal welfare. Last, but not least, the consequences of mega-farms for the environment were unclear. Kool *et al.* (2008) showed that the 'new mixed farm' in Grubbenvorst had advantages on a national scale with regard to sustainability, especially concerning ammonia, fine dust and stench. On the local scale, however, the new mixed farm would result in increased ammonia, fine dust and stench. So an independent study that probably was meant to ease the worries of local inhabitants actually lead to extra resistance.

9.3.4 Process manager

The reconstruction process was top-down. The government undertook action in the first place to restructure pig farming. Complex problems of spatial planning, environment, nature preservation, landscape, water and economics were added to the reconstruction process later on. County councils, municipalities and NGOs participated in the process. The provincial governments were responsible for composing reconstruction plans with the rural municipalities and social sectors. These plans translated the national government's starting points into actual measures and procedures tailored to the conditions in that particular area. This approach seemed to offer sufficient opportunities for social learning. However, the observed resistance among local people raises the question of whether the right persons were participating in the transition arena.

9.4 Theoretical reflection on the reconstruction process

The description above shows that the reconstruction processes in Markelo and Grubbevorst addressed complex problems in livestock farming in the Netherlands and were triggered by events revealing risks related to existing practices. The description also shows local resistance to the proposed reconstruction process. This section uses several theoretical inputs in an endeavour to explain such tension between a transition process intentionally moving toward more sustainable livestock farming and the social fabric in two municipalities.

9.4.1 New institutional economics: North and Aoki

According to the theory of new institutional economics, the starting point of (institutional) change is the individual and individuals inducing change can be labelled as 'entrepreneurs'. This is in agreement with the definition of entrepreneurship by 'old' economists who describe entrepreneurs as 'movers of the market' (Cantillon (1680-1734), Schumpeter (1883-1952), Kirzner 1973 cf. Van Praag, 1999), 'innovators' (Schumpeter cf. Van Praag, 1999) and 'discoverers of profit opportunities' (Van Praag, 1999). In the case under consideration about agricultural development areas, entrepreneurship hardly played a role (except for those entrepreneurs who wanted to move to agricultural development areas). Rather, the reconstruction process came into being in a 'top-down' way. Local municipalities and their farmers and citizens were confronted with the fact that their area was slated to become an agricultural development area. Yet in innovation processes, it is important that initiators seek stakeholders who have corresponding interests (Grin and Van de Graaf, 1996). In the case of the agricultural development areas, it is questionable whether the interests of the actors involved corresponded enough. It is assumed that the landscape and environment will improve in the expansion and intertwining areas, but the agricultural development areas suffer the consequences of this and local actors apparently objected to these consequences.

In terms of the new institutional economics, the transaction costs attached to building good relationships with local actors, and to including their interests in the change process induced by the national government, may be too high. Offering local actors a good alternative may help to lower these transaction costs. The theory of new institutional economics also emphasises the importance of 'institutional patches': carrying out a big change at once, might cause resistance, whereas changing in small steps might prevent this (more or less). Termeer and Van der Peet (2009) present a similar argument. They conceptualise emergent change as ongoing accommodations, adaptations and alternations that produce fundamental change. The new realities that began to become visible in Markelo and Grubbevorst could not be mastered overnight. A longer time span may in fact be required before small gains start to convince local actors that the process is heading in the right direction. Introducing a radical change in one single package may be beyond local capacities to adapt and manage the consequences of such a change.

Another assumption in new institutional economics is that institutional change occurs if an increasing number of people cease to comply with the rules. In the case of the agricultural development areas for livestock farming, the rules were not changed. On the contrary, the regime players in the national and provincial government imposed their plans on the local municipalities (and after accepting the plans, the municipalities more or less imposed the plan on their residents). In line with new institutional economics, institutional change is a necessary condition for a change process like the reconstruction process. As yet, however, nothing of the kind has happened in the reconstruction process. On the contrary, the reconstruction seems to have been 'pressed' into the existing system.

9.4.2 Social systems: Luhmann

According to Luhmann, social systems as well as organisations are characterised by a lack of observational capacity. Social systems are primarily driven by communications. Each social system has a unique code of communication, which can be observed by other social systems but can never be fully understood. As a consequence, it is impossible for organisations to 'read' other social systems completely, causing the lack of observational capacity mentioned earlier. A common feature of social systems is autopoiesis, a kind of organisational blindness, which is caused by the fact that a social system can only maintain itself while reproducing itself from its own elements, using its own structures and procedures. So, each organisation needs feelers to prevent them from getting 'trapped' in their own social system.

In the case of the agricultural development areas the process was directed by the federal and provincial governments. One might say that these institutions form a social system in which autopoiesis plays a role. This causes 'operational closure', which means that the boundary of the system might be open for observation and interaction with other social systems – in this case the local municipalities and their farmers and residents – but everything happening within the boundary has to be a product of the operations of the system itself. Thus local municipalities and their farmers and residents (who also form a social system) are framed according to the procedures and programmes, or the 'logic', of the social system formed by the federal and provincial government. These latter actors lacked observational capacity, which explains, in Luhmannian terms, why these governments were unable to take into account the specificities of other social systems in both localities. Nevertheless, the governments seemed to impose a generic recipe upon these different social systems, and even tried to incorporate them into their own system. This also makes clear why, according to social system theory, it is difficult to apply more or less the same reconstruction process (namely the one set up according to the procedure and programmes of the national and provincial governments) in different agricultural development areas: different social systems affect each area in a unique way. So, from the viewpoint of social systems theory, the more attention for the local circumstances the better. Cooperation with local actors is thus key. But who are the local actors? Project leaders or planners of a reconstruction process might easily overlook local social systems, and their endeavours may unintentionally even erode these social systems. One actually needs to be part of a social system – in this case, the network of local actors – to be able to fully 'read' it.

9.4.3 Strong and weak ties: Granovetter

Granovetter developed a theory in which the strength of interpersonal ties in small-scale interactions can be used to explain macro phenomena, such as diffusion of innovations, social mobility, the organisation of communities and social cohesion in general. He distinguishes 'weak ties' and 'strong ties'. The most important features of strong ties are the transfer of complex and tacit knowledge, high levels of trust, high relational embeddedness and the sharing of sensitive information. The most important features of weak ties are the transfer of simple explicit knowledge, lower levels of trust, low relational embeddedness and providing access to a greater amount and diversity of information.

For the analysis of the reconstruction process in livestock farming, this implies presenting outcomes as conditional on linkages among otherwise disconnected social networks via weak ties, for example, actors working at different scales, as was also put forward by Kool *et al.* (2008). On the other hand, strong ties are needed to exchange complex knowledge and to establish the trust that seems necessary for the multi-dimensional transformation of practices long established in livestock farming. Social networks theory makes the viability and effectiveness of the reconstruction process dependent on the balance between strong ties and weak ties.

The reconstruction process represents a complicated problem which may 'need' the features related to strong ties, while weak ties may also be needed to keep the people involved who are embedded in real life practices and who are most affected by the process. However, the reconstruction process started in a top-down way. At different levels in the process, strong ties were most important. Local networks were characterised by a high level of social embeddedness, both within the networks of public officials from the national and provincial government, who 'invented' the reconstruction process, and within the local municipalities and among their residents. This may have caused an unbalance between strong ties and weak ties – or weak ties may even have been missing – which could not be compensated by the advantages of strong ties.

A contrasting possibility is that the confrontation between the inventors of the reconstruction process and the local municipalities and their citizens actually created weak ties, which can be an advantage when it leads to mutual trust. The strong-weak tie framework can also point at strong ties, involving sharing of tacit knowledge and embedded in complex problem solving in specific agro-ecological environments, which may explain a certain level of resistance of local farmers and other local residents to 'newcomers' who have to be incorporated into existing networks. This is disadvantageous for mutual trust between actors embedded in a locality and actors bringing resources from outside the locality. Headlines in local and national newspapers and journals confirm this, for example, *Farmers in Reconstruction Areas Increasingly Diametrically Opposed to Each Other* (agricultural newspaper, September 2006) and *Own Pigs First* (local newspaper, November 2007, http://lexisnexis.academic.nl; Annex 1)

9.4.4 Doubt management: Weick

According to Weick, both too much doubt and a lack of doubt can inhibit the progress of (change) processes. The structure of sensemaking in such processes should not to be too tight and rigid. Because 'it takes a complex sensing system to register a complex environment', loosely coupled elements are necessary. No 'one best way of action' should be chosen at the beginning of a (change) process, but some room needs to be left for randomness and trial and error. In the case of the agricultural development areas, the designers of the reconstruction process seem to work with a single solution, based on a combination of technical and political considerations. A result of this was that the intended change process was presented and implemented in a top-down mode. This reduced the 'room for manoeuvre' for the local actors.

The description of the reconstruction process in livestock farming shows that, despite the best of intentions of government officials in shaping a new way of livestock farming, the real nitty-gritty change of practices and tensions surfaced when down-to-earth matters arose of what to do in actual practical terms in specific localities. As long as the reconstruction process was far away and consisted merely of plans on paper, local actors were not incited to move. Only at the point when the reconstruction plans became real did local actors mobilise. At that point, however, the process of operational closure had already progressed so far that little room was left for doubt or correction. This resulted in alienation, which may explain the resistance of local actors. A prejudiced point of view about the outcome, of a reconstruction process, for example, might cause matters to be overlooked that are nonetheless important for specific regions. According to Klerkx *et al.* (2010), agricultural innovation policies should, instead of aiming to fully plan and control innovation, foster the emergence of instruments that enable adaptive innovation management.

9.4.5 Power and discipline: Foucault

Foucault defines a discourse as a practice of strategic games in which realities are constructed. The reconstruction process aimed for 'reconstructed' rural areas, with an improved environment and more attractive landscape. This was the reality of the people who 'invented' the reconstruction process. However, they forgot another reality, namely that the reconstruction process had consequences for local communities and the 'not in my backyard' proponents. Certain knowledge about (the advantages of) the reconstruction process seems to have been accepted or 'embraced', while other knowledge or unintentional effects (e.g. for the local communities) were excluded. This can be read in many of the headlines in Annex 1:

- Commotion about blocks of pig farms: fear for stench, trucks with piglets and big buildings
- Reconstruction law opens door to pig flat, reorganisation of rural areas has unintentional effect: small family farms disappear
- Doctors protest against pig and chicken factories, agropark: 'so much fine dust and they know how great the risks are'

In Markelo and Grubbenvorst only a very small majority was needed to get the reconstruction process plans approved by the municipalities. Farmers who wanted to participate in family farm plus (Markelo) and the new mixed farm (Grubbenvorst) were nonetheless given permission to start permit procedures. This indicates that the interests of local farmers and other local residents were more or less neglected.

Foucault's theory would focus attention on the mechanisms of disciplining, exclusion and subjection visible in the way the national and local governments aligned to reconstruct livestock farming.

9.4.6 Materiality: Benton

According to Benton, intentional processes of change are limited by materiality, and thus by natural limits and material conditions. Innovation processes often are too directed at transformation (technological processes), although environmental limits and the low level of material malleability are important characteristics of the agricultural labour process. Nature's dynamics operate in a stratum distinct from society, as revealed by the outbreak of classic swine fever, but also affect society's capability to respond to processes in nature with are largely independent from human action and social organisation. The emphasis on the potency of technology in livestock farming may have concealed that an agricultural labour process is perhaps best conceptualised as an eco-regulatory structure, hence focusing on the conditions for animal growth rather than on transforming animals into meat products.

The health risks attached to animal diseases induced the establishment of a hierarchical social organisation. It is questionable, however, whether an exclusive focus on such an institutional modality is the right answer to independent biological processes. Moreover, the reconstruction process modified the current practice of eco-regulation of intensive livestock farming only to a limited extent; the keeping of pigs and piglets did not change, only their spatial distribution changed, leading to even more pigs in the agricultural development areas.

With regard to the spatial distribution, however, the reconstruction process may be a well intended attempt to integrate environmental and social processes. After all, the agricultural development areas could improve the environment and the landscape, and offer farmers the opportunity to move their farms to a location where they could extend their farms without affecting vulnerable nature. The inventors of the reconstruction, however, underestimated the reaction of the local communities. So environmental and social processes were integrated with regard to the farmers, but not with regard to the people living in the area (and the animals kept in intensive livestock farming).

Kool *et al.* (2008) examined the implications of the 'new mixed farm' in Grubbevorst. The results showed that the new mixed farm has advantages on a national scale in terms of sustainability, especially concerning emissions of ammonia, fine dust and stench. On a local scale, however, the new mixed farm would lead to increased emissions of ammonia, fine dust and stench. These results were perhaps disappointing for the advocates of the new mixed farm. According Benton's theory, however, it does not matter whether the results were positive or negative. Mere calculations, even if they are favourable, cannot take away the concerns of local actors because they are unrelated to local actors' motives for rejecting the new mixed farm. Albers *et al.* (2006) describe all possible consequences of the new mixed farm in a rather neutral way. In this publication not a word is mentioned about the consequences of the new mixed farm for local actors.

9.5 Synthesis

An important feature of the reconstruction process of livestock farming is that its design and implementation came into being in a top-down way. Local municipalities and their residents were confronted with the fact that their area was slated to become an agricultural development area. This did not strongly affect them as long as it was only a matter of plans on paper. But as soon as these plans became more real, resistance rose. Applying different theoretical perspectives to the reconstruction process led us to identify a number of competing explanations for why the process did not work out as expected.

According to the new institutional economics, a regime can change its direction if an increasing number of people resist current rules and arrangements. If a transition process comes into being in an overly top-down manner, it may lead to collective action against the transition by people who are not involved in the process. This is not as likely to happen if these people are involved from the beginning and if the change aimed for can be established gradually to keep the transaction costs under control.

A complication, according to the social systems theory of Luhmann, is that it is difficult to ascertain who exactly are the local actors needed to make the transition process successful, because one has to be part of a social system to fully understand it. The reconstruction process presented in this chapter may be better explained by looking into the autopoiesis within a variety of social systems as well as into the balance between strong links, found in the affected local communities and in the professional associations implanting the plans, and weak links. Knowledge may remain implicit and a connection with other social systems with probably a completely different and illuminating view on the reconstruction process may be omitted.

According to Foucault, limiting involvement in the reconstruction process to a selection of actors may enforce a certain discourse or presupposition – for example, based on technical and economic calculation–, which can be expected to lead to resistance at a local level. This may exclude room for 'reasonable doubt', which is important according to Weick's theory of doubt management. It also fits in the theory of Benton, that not only are technological and transformation processes important for realising change, but also less malleable natural processes and social structures should be taken into account. After all, even if the 'hardware' is more or less comparable in the different agricultural development areas, the reconstruction process may differ completely between these areas because of natural circumstances and social structures. Knowledge of the local social circumstances is important in each agricultural development area where the reconstruction process takes place. Each 'transition manager' designing a transition process must be able to answer the question: 'Who do I need, when and in what way?'

References

Albers, K., L. Lamers and H. Hullenbroeck (2006) Startnotitie Nieuw Gemengd Bedrijf: Gemeente Horst aan de Maas (Knowhouse Fresh Innovations). Arcadis, Amersfoort, the Netherlands.

Boonstra, F.G., W. Kuindersma, H. Bleumink, S. De Boer and A.M.E. Groot (2007) Van varkenspest tot integrale gebiedsontwikkeling: evaluatie van de reconstructie zandgebieden. Alterra Report No. 1441, Alterra, Wageningen, the Netherlands.

Geels, F.W. (2002) Technological transitions as evolutionary reconfiguration processes: a multi-level perspective and a case-study. Research Policy 31: 1257-74.

Grin, J. and H. Van de Graaf (1996) Technology assessment as learning. Science Technology and Human Values 21 (1): 72-99.

Klerkx, L., N. Aarts and C. Leeuwis (2010) Adaptive management in agricultural innovation systems: the interactions between innovation networks and their environment. Agricultural Systems 103: 390-400.

Kool, A., I. Eijck and H. Blonk (2008) Nieuw Gemengd Bedrijf: duurzaam en innovatief? Blonk Milieu Advies, Gouda, the Netherlands.

Rotmans, J. (2003) Transitiemanagement: sleutel voor een duurzame samenleving. Van Gorcum, Assen, the Netherlands.

Rotmans, J., R. Kemp and M.B.A. Van Asselt (2001) More evolution than revolution: transition management in public policy. Foresight 3 (1): 15-31.

Termeer, C.J.A.M. and G. Van der Peet (2009) Governmental strategies and sustainable transitions: monitoring systems for the prevention of animal disease. In: K.J. Poppe, C.J.A.M. Termeer and M. Slingerland (eds.). Transitions Towards Sustainable Agriculture and Food Chains in Peri-Urban Areas. Wageningen Academic Publishers, Wageningen, the Netherlands, pp. 253-271.

Van Praag, C.M. (1999) Some classic views on entrepreneurship. The Economist 147 (3): 311-35.

VROM (2003) De reconstructiewet: reconstructie en ruimtelijke ordening in de praktijk. Ministerie van Volkshuisvesting, Ruimtelijke Ordening en Milieubeheer, Den Haag, the Netherlands.

Website

LexisNexis: http://lexisnexis.academic.nl.

Annex 1. Examples of negative headlines about the reconstruction process and agricultural development areas in Dutch journals and newspapers between April 2006 and October 2008 (taken from: http://lexisnexis.academic.nl).

Headline	Kind of source	Background
Farmers angry about mega farm	Provincial newspaper, April 2006	Local pig farmers are angry that ground they wanted to buy has been sold to a pig farmer from 'outside'
Reconstruction slower than planned	Agricultural newspaper, May 2006	The implementation of reconstruction plans is taking more time than expected because municipalities fear a pig invasion and farmers who want to move to agricultural development areas think that a lot more is allowed in those areas than elsewhere, which causes permit procedures to take a lot of extra time
No new farms in agricultural development area after all	Local newspaper, June 2006	Municipality wanted to allow newcomers in the agricultural development area. This met resistance from local farmers. A new alderman changed the plans
Support key to success of pig cluster	Agricultural newspaper, July 2006	Recommendation of researcher after failure of another agricultural development area due to lack of support of local actors
Farmers in reconstruction areas increasingly diametrically opposed to each other	Agricultural newspaper, September 2006	Dairy farmers in agricultural development areas are afraid that their activities will get stuck because of the activities of pig farmers
Walls of resistance	Farmers' journal, June, 2007	Discussion of obstacles to the reconstruction process, such as resistance of farmers and citizens, increasing costs and permit procedures
Practice moderates strive for reconstruction	Farmers' journal, June 2007	Complaint that in the reconstruction process more piles of papers are replaced than farms
Province wants to continue with plans in pig regions	Local newspaper, July 2007	Provincial government, municipalities and district water board want to continue with 26 agricultural development areas in the Province of Overijssel, despite the resistance of area inhabitants against newly built big pig farms in their region
Commotion about blocks of pig farms: fear for stench, a lot of trucks with piglets and big buildings	National newspaper, July 2007	See headline

Annex 1. Continued.

Headline	Kind of source	Background
Reconstruction law opens door to pig flat, reorganisation of the rural area has unintentional effect: small family farms disappear	National newspaper, July 2007	See headline
Pig flats do not belong in the Netherlands	Farmers' journal, September, 2007	See headline
A agricultural development area is a lot of pink, hidden in the green	Provincial newspaper, September 2007	Agricultural development areas are needed because people do not want to live in the stench of pig farms, but they do want to buy cheap pork
Mega-farm is obstacle in the reconstruction process	Agricultural newspaper, November, 2007	A lot of commotion arisen because the reconstruction plans are associated with mega-farms
Fear for mega-paralyses	Farmers' journal, November 2007	Municipalities are so afraid of mega-farms in their area that they tend to keep out all activities related to intensive livestock farming
Own pigs first	Local newspaper, November 2007	Although it is against the intention of the reconstruction process, many municipalities choose a 'not in my backyard' approach in the sense that the agricultural development areas are only for their 'own' farmers who want to farm or expand their farm operations. This is called the 'own pigs first' approach
Doctor's protest pig and chicken factories, Agropark: 'so much fine dust and they know how big the risks are'	National newspaper, December 2007	See headline
Grubbenvorst Christmas theme – under the spell of the mega-farm	Farmers' journal, December 2007	See headline

Annex 1. Continued.

Headline	Kind of source	Background
Protest against mega-farm expansion continues	Dutch newspaper, January 2008	See headline
Thousands of signatures against pig invasion	Local newspaper, March 2008	See headline
Let them all piss off	National newspaper, May 2008	See headline
A lot of resistance from society	Farmers' journal, June 2008	See headline
Wrestling with mega-farms	Farmers' journal, August 2008	See headline

Chapter 10

Seed provision in developing economies: converting business models

Rolien C. Wiersinga, Derek Eaton and Myrtille Danse

10.1 Introduction

Transition management is focused on complex problems in society and tries to organise processes to find long-term sustainable solutions (Rotmans *et al.*, 2005). This chapter describes a case of transition aimed at increasing smallholders' income through higher crop yields in South-East Asia. The strategy was to introduce improved seeds and make these accessible to smallholder farmers, facilitating their access to higher quality local vegetable market segments and thus leading to improved earnings. The case described has succeeded in increasing smallholders' incomes and expanding vegetable seed markets.

This transition in the Asian tropical horticultural sector is conceptualised as the result of private companies tailoring technologically innovative products and processes to markets consisting of many smallholders. This implies a balancing of three A's: 'affordability', 'access' and 'availability'. The approach relies on dovetailing quality products and processes with the demands and needs of customers who have low incomes and who face specific constraints, particularly in relation to the combination of traceability, quality, product development and added value for smallholder farmers. Firms must be able to manage a variety of business models, market strategies and modes of chain coordination to offer the flexibility needed in markets with many smallholders. Consideration of smallholder farmers in developing countries as a potential market for innovative technologies is elaborated by the Base of the Pyramid (BoP) ideas of Prahalad (2004). Targeting smallholder farmers as resilient and creative entrepreneurs, conscious of the value of new technological opportunities, challenges the current business models of firms as well as (public) strategies for technology transfer (London and Hart, 2004). Nevertheless, the development of such business models is expected to benefit those private technology providers that prove capable of tapping into the markets of rural producers with less purchasing power (Hart and Sharma, 2004; Prahalad, 2004; Danse and Vellema, 2007).

In retrospect, this case can be viewed as a successful example of a BoP approach. One of the companies in particular, East West Seeds (EWS), perceived a market opportunity among smallholder farmers already undertaking some form of commercial horticulture. The company's vision was to bring to South-East Asian farmers high-quality vegetable seeds at an affordable price. As these farmers have relatively modest assets (land, capital, low-quality seeds), EWS developed suitable small-scale production formats and invested considerably in demonstration and transfer of knowledge of improved cultivation practices. The vegetable varieties sold can be seen as part of a technological package that offers best results in combination with specific crop fertilisation, irrigation and pest management practices. The company managed to become socially embedded, which is a key factor in international business strategies aimed at obtaining access to the BoP market (London and Hart, 2004). Besides that, the EWS-initiated transition process led to a new unique product/service package that met the needs of low-income groups. Moving to South-East Asian countries, setting up local entities and establishing local distribution centres made the company's hybrid seeds available to the growers and contributed to their adaptation to local wishes. Capacity building, training local breeders and establishing participative experiments enabled EWS's knowledge to be translated to local mechanisms and traditions. Offering various packages with smaller and larger quantities of seed improved the affordability of the product, which unlocked the BoP demand for improved seed.

An open question is whether the success of the improved seed materials will create the need for a 'sustainability transition' in the future. If so, will seed companies be as entrepreneurial as before? Or will the public sector or civil society have to assume a 'clean-up role', to put it bluntly. The seed companies have introduced hybrid vegetable seeds that involve intensive production, using higher doses of synthetic fertilisers and pesticides (among other inputs), with little immediate concern for the positive or negative environmental consequences (externalities). Though recognising this, the current analysis does not examine the environmental effects of the innovations.

10.2 Reconstruction of the process

10.2.1 How it all started

In the second half of the 20th century, the Green Revolution drastically transformed Asia's rural areas. The introduction of improved plant varieties of staple crops generated, in general terms, in an improved food security and increased rural incomes, although the effects differed for areas and groups. More recently, policies and regulations that aim to secure access and re-distribution of land to smallholders created new conditions for fighting poverty in countries with high levels of rural poverty, such as Vietnam and China. Starting point of this chapter is a discussion on additional policy changes and strategies needed to maintain the momentum of the reductions in rural poverty and social inequalities, to improve food security and to provide access to safe and nutritional foods.

In Asia some 250 million farmers grow vegetables. These activities are generally labour-intensive and undertaken for the most part by smallholders. Due to its labour-intensive nature and association with higher value-added agricultural products, vegetable cultivation appears to have made a significant contribution to farmers' incomes and poverty reduction. But little research has documented this process.

Vegetable production has been common in China for a very long time. Due to the migration of Chinese people in the past, the vegetable production system was brought to South-East Asia as well. Vegetable seeds were commonly bought and sold by the Chinese in Asia until the communist upheaval disrupted this seed trading system. As a result of this change in political system, the seed companies were nationalised and the quality of vegetable seeds diminished. After the 1950s the traditional supply of Chinese seeds to other countries of Asia became less common than before (EWS, personal communication). Nowadays, the types of vegetables cultivated and consumed varies per country. For example, hot pepper culture is common in Indonesia and Thailand, but not in the Philippines.

The decreased Chinese seed trade in the Asian market provided an opportunity for non-Chinese seed companies. A Dutch seed company displayed interest in the region, due to the large Asian population with vegetables in their main diet. It set up a local vegetable seed company, East West Seeds. The Dutch seed company was part of the development of F1 hybrids in Europe from 1950 to 1980, which also took place in the United States and Japan. The president of this Dutch company, Simon Groot, was aware of the value of the F1 technology for increasing the genetic quality of crops and therefore for increasing production and incomes. The change of the Chinese political system and the imbalance in the vegetable seed system created a situation which was adaptive and receptive to innovations like improved vegetable seed materials.

10.2.2 The change process envisioned

The system innovation started 25-30 years ago when vegetable seed companies (led by EWS) observed an opportunity to breed improved varieties of vegetables crops that were widely consumed in South-East Asia. These crop varieties had received very little attention from public agricultural research organisations, since they devoted their efforts primarily to staple crops such as rice and maize. Hybridisation would likely lead to varieties displaying considerable improvements in agronomic and post-harvest characteristics.

In 1981 Simon Groot went to South-East Asia to explore the region, together with the president of an American seed company and a tropical vegetable expert. Vegetable plant breeding was not yet being done in South-East Asia (EWS, 2002). Farmers there saved seeds of local vegetable crops and vegetable seeds were also imported into the region. The varieties used were either producing low yields or not adapted to the region's environment and consumer wishes or both. In 1982, Groot founded East West Seeds jointly with a local seed trading business. EWS was founded to breed local varieties locally for the local market. The company's strategy was based on two major beliefs. The first one was that three to five years of work would need to be invested before breeding could yield results, and this process could not be rushed. The second belief was the high value of knowledge of local production conditions and local consumer preferences. The company saw this knowledge as of paramount importance to success, as farmers only buy seed after witnessing their performance. Besides that, word of mouth was believed to be the only effective advertisement.

The choice to start the venture in the Philippines was mostly practical. It was necessary to gather a lot of local knowledge about the crops, diets and culture, so it was useful to be able to speak the local language. In the Philippines, many people speak English fluently and it was therefore anticipated that in this country the required knowledge could be gathered faster than in neighbouring countries. Breeding work in the Philippines started in 1983 on five hectares with a seed processing plant and offices. In 1986 the first hybrids and improved open pollinated selections were introduced in the country. While the venture in the Philippines was getting started, the company opened a second office in Thailand in 1984 with a farmer-oriented tobacco leaf company diversifying into vegetable seed production. 'It was difficult to interpret the scant information, put aside our own biased opinions and understand what local farmers valued' (EWS, 2002). In 1986 the first EWS varieties were released in Thailand and a year later the first hybrid variety. In 1990 EWS started, together with another Dutch seed company, breeding vegetables in Indonesia. Two years later the first hybrid was released. In 1995 the regional head of the Asian Vegetable Research and Development Center in Thailand requested EWS to start a breeding programme in Vietnam. A year later the first (jointly developed) hybrid was released and in 1997 the joint venture was registered. In each of these four countries EWS was a national pioneer in commercial vegetable plant breeding. In 2001 variety testing started in Africa. What EWS has achieved up to today was not foreseen by the company when it started in the Philippines. The growth of EWS in South-East Asia was a step-by-step process.

Reflecting on its success, EWS states the paramount importance of being consistent and keeping focused. The overall focus of the company is to increase the income of smallholder farmers. At first, this was accomplished primarily by increasing production per hectare; now a broader aim is more constant production year-round.

10.2.3 Key success factors for adopting new technology

As a result of improved local vegetable varieties, farmers in general now earn a higher income than before (Eaton and Wiersinga, 2009). Clearly the structure of the (horticultural) system has changed fundamentally, as well as the relations among the different stakeholders in the system. The improved varieties have mostly been offered as part of a package of technologies including plastic mulch, fertiliser, pesticides and staking. Other changes at the household level that often go hand in hand with the adoption of improved varieties are changes in labour tasks within the household and different trade channels both for inputs and for the output.

The importance of developing trust is quite considerable in the case of commercial seed purchases. When a farmer buys seed, there are very few possibilities for judging their quality, in terms of their potential productivity or the characteristics of the resulting produce. This will not become apparent until many weeks or months later, when the seed seller could easily be long gone. Even after the season has transpired, an assessment of the seed's performance is still complicated by other confounding factors. Thus, it should be apparent that the decision to purchase seed from a relatively unknown outsider would entail considerable risks in the eyes of the farmer. But of course, demonstration effects might work quickly and effectively.

The factors affecting adoption can be clustered around varietal, farm, farmer and more institutional factors. As is to be expected, farmers concentrate on varietal characteristics, in particular, on improved yields, fruits that reflect market demands, and ease of maintenance in the field. Farmers mention higher yields and fruits that better reflect market demands as the main factors instigating adoption. A commonly-cited reason for farmers not to adopt hybrids is distrust in the quality of the seed, as outputs may be disappointing in terms of yield and produce quality.

Whether the potential positive characteristics of hybrids are fully reflected in the produce depends on the environment in which they are grown, that is, farm-level characteristics. The variety needs to be cultivated in a geographically suitable location, and a proper road and adequate proximity to the market is required for produce to arrive fresh at the point of sale.

In terms of famer-level factors, the willingness of farmers to bear risk as well as their agronomic expertise affect the decision to adopt hybrid varieties and innovations in general. Characteristics such as age and education could quite plausibly influence openness towards new technologies. Adoption may also be correlated with factors such as access to credit or a preference for cultivating traditional varieties for local consumption. In a number of cases, lack of capital is a reason for not adopting a hybrid variety. Capital may be required to buy plastic mulch, fertilisers and pesticides, in addition to the seed.

In terms of institutional and market factors, demand for (the harvested products of) improved varieties is the key reason for selecting them. But it is difficult to completely separate the varietal, farm and farmer-level factors mentioned above from institutional-level factors. For example, access to credit is referred to as a farmer-level factor, but this relates not only to individuals but also to the institutions and mechanisms for providing farmer credit. Indeed, overall the most important factor constraining farmers from adoption appears to be the necessity to purchase not only seeds but also related inputs such as agrochemicals and plastic. In some instances, the seed company's agent may extend some credit, though not all farmers appear able to access this. This suggests that the attention of policy-makers and seed companies needs to be directed towards innovative solutions for overcoming credit constraints and towards associated measures to help prevent farmers from getting caught in a debt trap (Eaton and Wiersinga, 2009).

10.2.4 Private-sector driven institutional change

In this case, the entrepreneurial seed companies can to some extent be seen as conscious initiators or managers of the transition. They worked actively to develop their markets and to change the cultivation practices of vegetable farmers. When EWS started to breed local vegetable seeds, it turned out that farmers often did not know how they wanted the vegetables to be improved. Often the traders had the clearest view of what was important, for example, a longer shelf life. The government played a minimal role. This is quite different from the Netherlands, where the EWS pioneers were from. In the Netherlands the government and research institutes are closely involved in the development of the horticultural sector.

In the Philippines the government did play a role. First, a public-private partnership provided knowledge to smallholder farmers and worked with them to increase their skills in high-income vegetable production. Farmers received technology packages (including seeds, trellises and plastic mulch) along with intensive training and technical support from the private sector, sponsored by the government. Fields of these farmers functioned as demonstration plots for other farmers (EWS, 2002). Second, development of the hybrid eggplant variety 'Casino' was subsidised by the government in the Philippines.

In Vietnam the government also played a role with its programme to develop horticulture in the 1999-2010 period. The Vietnamese government encouraged the use of improved seeds. In some villages in South Vietnam, farmers attended training organised by the government in cooperation with companies on how to use new varieties and related farming technologies. Through these efforts, farmers felt supported in their choice to adopt the new varieties.

Perhaps the most defining characteristic of the process of diffusing hybrid vegetable cultivation technology was the leading role played by private-sector breeding companies. In some cases, farmers were previously cultivating varieties developed by public-sector organisations, but elsewhere the introduction of private-sector hybrids represented a fundamental shift in agricultural systems. The companies not only developed the varieties, but also undertook active marketing and extension efforts to demonstrate to farmers the performance of the technology. The company or its agents played a role in introducing the technology to farmers, and usually more so, as they often arranged demonstration plots and even provided credit for seed or other inputs. Promotion of hybrids, either by the seed company or by the government, stimulated adoption. This suggests the importance of considering innovative ways of reaching more farmers with information and demonstrations.

10.3 Transition reflection

This case can best be seen as an example of market-driven technological change and economic growth in primary agricultural production. In many ways, the process of technological change was similar to that which occurred with the introduction of high-yielding varieties of staple food crops during the Green Revolution in Asia. But there are some salient differences. The process of change in the vegetable sector deals with a changing market regime. EWS initiated the development of new forms of self-regulation and began to define alternative schemes for introducing seed materials in an informal market. In this case, the institutional arrangements regulating intellectual property rights in technological innovations have less value than trust and management of social embeddededness in the community. EWS stayed ahead of those who were reproducing its varieties without a licence by developing and introducing a steady flow of small variety improvements, especially for genetic disease tolerance. In general terms, EWS and its Asian partners have accepted the likelihood that roughly 20% of the total market will be served by copies.

The most important innovative aspect of the business model developed by EWS is perhaps its investment in services and products provided to smallholder farmers. Following essentially a BoP strategy, a handful of companies catalysed a broad and far-reaching change in technology, in economic production and relations.

Further, this case seems to be a sectoral system of innovation, in the sense that a group of firms is involved that is active in developing and making a sector's product (vegetable seeds) and in generating and utilising a sector's technology (breeding) (Breschi and Malerba, 1997). Geels (2004) states that the sectoral system pays less attention to diffusion and use of technology than to the development of knowledge. This does not apply to this case, however, as the diffusion of the improved seeds is clearly of high commercial importance to the seed companies, which also offer farmers knowledge about the improved varieties.

Rotmans *et al.* (2005) define system innovation as innovations that go beyond an organisation and which radically change relations among the companies, organisations and individuals in the system. The case of improved vegetable varieties did radically change these relations. For example, farmers often switched input suppliers as a result of adopting improved varieties. Some traders provided credit in addition to seeds or offered a guaranteed market for outputs.

At the regime level, this case shows changes in the property rights regime concerning genetic resources. The innovation code is locked in hybrid varieties by the seed companies. As a result, farmers have to buy the improved seeds every year and cannot multiply the seeds themselves. Whether improved seeds have a political effect built into them or, using the notion of Latour (1992), a 'script', is still under discussion. As market demand among farmers increases, seed companies must step up production. This is already induced by introducing new institutional arrangements, such as larger buying firms contracting smallholder farmers to grow vegetables, which can be viewed as relatively novel in South-East Asia's cultural environment. Under these contractual conditions, the genetic material becomes private property.

At the meso level, one important transition for the farmers converting to the use of hybrid vegetable seed was the increased or enhanced degree of commercialisation of farm production. Due to new market requirements, awareness-raising among farmers was needed to stimulate their willingness to invest in the purchase of inputs for crop production. These inputs helped them to obtain higher yields and, as such, created opportunities for sales of the resulting surplus. Prior to the adoption of the new technology, this marketable surplus was considerably less. The shift to purchased inputs involves a change in farmer habits and customs. Previously, seeds were saved and exchanged among farmers. In many communities, there were certain farmers who were specialised in the management of genetic resources, including storage of a wider range of seed and ongoing selection activities. In the process of disseminating hybrid seeds, seed companies, targeted identified individual farmers, who the companies considered to be able to demonstrate the tangible benefits of new planting material. In this way they were able to (slowly) persuade farmers to experiment, first on a limited basis, leading incrementally to a change in their habits. For many farmers, this probably also involved the evolution of trust in the marketing agents of the seed company and, in more abstract terms, in the market process in general. The specialisation in vegetable production and increased commercialisation of the local economy creates new relationships of economic dependence.

In terms of experiments, this case consists of a technological experiment (breeding), a market experiment (South-East Asia and BoP as something with growth potential), and an institutional experiment (involving the development of commercial trade in seed). For each type of experiment, EWS closely cooperated with locally established institutes, such as local universities, research institutes, seed companies, trading companies and the government.

Innovation was possible in producing hybrid varieties of local vegetables for local markets, as companies made strategic investments in creating niches. Introducing new technologies is a strategic choice that the companies made to gain a competitive advantage. This competitive and strategic game between firms opened up the regime. At the micro level, the transition germinated in part owing to the experimental demonstration activities of the seed companies. Overall, the experience provided evidence that small-scale, relatively cash-poor farmers could be persuaded to purchase commercial hybrid varieties.

In niches, as opposed to regimes, rules are not stable and do not yet have a strict structuring effect. Geels and Raven (2006) developed a framework helpful in describing the dynamics of niche development trajectories (Figure 10.1). First, local practices are developed starting with experiments. In the EWS case, this occurred in the Philippines at a company demonstration farm. After the first successes of locally bred hybrid vegetables, pilots were also started in other Asian countries, such as Thailand, Indonesia and Vietnam. This occurred in cooperation with local bodies. The concept of learning by doing and learning effects as part of transition management (Rotmans *et al.*, 2005) does apply to how the seed companies operated in this case. The dissemination of hybrid seeds provided further learning. Farmers

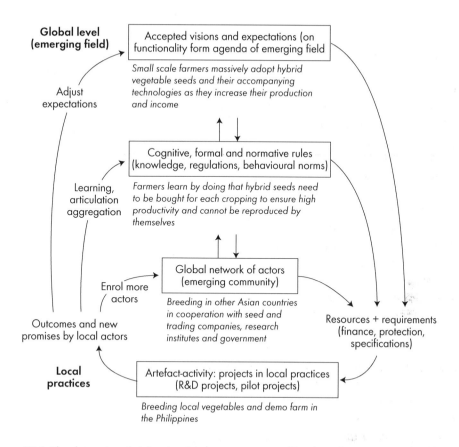

Global level (emerging field)

Accepted visions and expectations (on functionality form agenda of emerging field

Small scale farmers massively adopt hybrid vegetable seeds and their accompanying technologies as they increase their production and income

Adjust expectations

Cognitive, formal and normative rules (knowledge, regulations, behavioural norms)

Farmers learn by doing that hybrid seeds need to be bought for each cropping to ensure high productivity and cannot be reproduced by themselves

Learning, articulation aggregation

Global network of actors (emerging community)

Breeding in other Asian countries in cooperation with seed and trading companies, research institutes and government

Enrol more actors

Outcomes and new promises by local actors

Resources + requirements (finance, protection, specifications)

Local practices

Artefact-activity: projects in local practices (R&D projects, pilot projects)

Breeding local vegetables and demo farm in the Philippines

Figure 10.1. The dynamics of niche development trajectories (Geels and Raven, 2006), with added information in italics from the EWS case.

first had to get used to the new type of seed. The farmers were used to reproducing seeds themselves. However, they saw that the quality was markedly diminished when hybrid seeds were reproduced in the traditional way. It is inherent to hybrid seeds that they must be bought anew every season. This was soon accepted by the farmers, as they saw the large increase in productivity and their own income when using hybrid seeds.

10.4 Theoretical reflections

10.4.1 New institutional economics: North and Aoki

Whereas the transition literature focuses on managed change, NIE provides a framework to help us to understand autonomous change in economic systems. In this case, the role played by East West Seeds stands out, as this was one of the first companies to invest in breeding improved varieties of Asian vegetables. The expectation on the part of EWS that this was likely to be a profitable investment represents an example of this company's specific strategic management capabilities. While this concept, strictly speaking, comes more from management studies literature, its implications for heterogeneity among firms and innovation in business practices has been fully incorporated into NIE (Winter, 1984; Teece *et al.*, 2001; Langlois, 1992) although North does not appear to have made as much use of this stream of literature in his discussion of entrepreneurs.

Key aspects of the pattern of growth observed in the adoption of the new seed technology can also be explained within the NIE framework. The now-classic S-shaped innovation adoption curve can be viewed through the NIE lens as representing a first gradual, then accelerating and finally plateauing, change in habits and norms, in this case, of farmers cultivating vegetables. In the framework presented by Aoki (2001), this involves a move from one type of institutional equilibrium to another. At first, farmers were generally cultivating vegetables for their own consumption and some local sales, using farmer seed systems for genetic resources. The final plateau represents a new equilibrium in which (adopting) farmers have come to trust and rely on the practice of commercial seed purchases. This change began slowly, as EWS began with the first varieties in certain locations, also investing in demonstration and extension services. EWS agents sought to identify the few 'entrepreneurs' among local farmers who had an interest in experimenting, on a limited basis, with the new technology. Where success was demonstrated, other farmers gained interest and considered the uncertainty (transaction cost or risk) associated with this new technology. A large part of the uncertainty arose from purchasing seed that was 'foreign' and not self-managed, as well as from the need for other inputs that had to be bought up front at the start of the season and carrying the risk of possible crop failure or inadequate market demand. As more and more farmers successfully tried this new system, the uncertainty was reduced, leading to an acceleration in adoption. In so doing, farmers who previously would have been very reluctant to pay, say, a tenth of the value of their production, up front for seed, changed their beliefs and habits. In some instances, the companies may have reduced some of the risk by offering credit for input purchases, although this does not seem to have been widespread.

This institutional change is self-reinforcing, in the manner stressed by Aoki, because of the technological nature of hybrid varieties. Farmers not only shifted to the purchase of commercial vegetable seed, they also relinquished their management of genetic resources (saving seeds for the next season). With hybrid varieties, the reduced performance of saved seed strengthens the incentive to purchase new seed every cultivation season. This means that this change in technology and farmer practices, i.e. in the institutional structure of vegetable production, was not contingent on major changes in formal rules, such as in the legal regime. It also helps explain the faster growth, in terms of both productivity and scale, of vegetable crops that could be successfully hybridised, in contrast to others, such as lettuces, which could not. A similar transition for those crops is generally more dependent on change in formal institutions, such as the introduction of plant variety protection, possibly with legal restrictions on seed saving.

The NIE emphasis on path dependency in institutional change is also relevant in explaining this case. In relinquishing their own management of vegetable genetic resources (in exchange for increased profits), farmers engage in a process that is difficult to reverse. Once most farmers in a given locality have shifted to commercial hybrid varieties, the local varieties previously used may no longer be available. This may turn out to be a disadvantage, for instance, if a large number of farmers encounters a serious problem of reduced pest resistance in the modern varieties and they are unable to revert back to their previously cultivated varieties. While these may not have been as productive, they might generally have offered more consistent, albeit lower, overall performance. Thus, the initial process of institutional change directs farmers along a specific path.

10.4.2 Social systems: Luhmann

The innovation of locally developed hybrid vegetable seeds for the local market was developed in a small social system. The innovation started within seed companies that often were not collaborating with one another. A company like East West Seeds can be seen as an operationally closed social system strongly focused on breeding. Breeding is a self-reproducing process in this social system. All activities the company performs are focused on its own operative system. The company experiences that its system is working and is therefore driven to continue it. The positive feedback ensures the maintenance of the system. If the social system were to interact with other systems, the outcome is unsure and therefore not preferred.

This case reflects Luhmann's idea that no centre is needed to find a balance between self-reproduction and external relationships that allow social systems to innovate. What is needed is a set of interdependent yet operationally closed social systems. However, further innovations might occur at an earlier stage if companies know that their innovations are protected, for example, by patents and property rights. In organising a system of patents and property rights, there is a potential role for the public sector, which in this case has not (yet) had much involvement in this transition. East West Seeds has requested the public sector to better support the horticultural sector which shows that the company is trying to look beyond its own capacity and responsibilities.

Because societies are polycentric and radically differentiated, Luhmann claims that there is no guarantee for one *recipe* to be successful in the many different systems. Due to different situations and cultures, intentions and outcomes are difficult to predict. It is necessary to know the local situation in terms of politics, economy and cultural aspects. Luhmann warns against having too many ambitions with perfect project designs. Due to the many differences, it is vital not to try to block ideas that might come up at a later stage. This case shows that the recipe of local breeding of local varieties for the local market has been successful in a number of countries and in a number of different situations. EWS has an office in every country in which it sells seeds. However, clearly, in some locations the 'recipe' has been more successful than in others. For example, the role of extension agents and seed traders often seems to be critical to the successful diffusion of hybrid seed adoption. The assumption of well-functioning extension departments and traders working for the benefit of the seed company may have been part of the autopoiesis of the seed companies. In some countries or locations this assumption did not stand strong; there more attention for local conditions would probably have led to a more successful diffusion of the adoption of hybrids.

An interesting link to this theory is the Bottom of the Pyramid (BoP) concept. BoP claims that a certain product or service can be successful in different settings. The level of success in new environments depends on whether those newly selected environments have some shared elements. Due to their individual autopoiesis, the environments are not similar in every way, but can be similar in certain regards, such as having a critical mass of smallholder farmers looking forward to earning some additional income, a substantial vegetable market and existing trading systems.

10.4.3 Strong and weak ties: Granovetter

Granovetter states that the strength of interpersonal ties in small-scale interaction helps to explain macro phenomena. Tie strength can explain interactions between actors and between different systems. Weak ties are more likely to link members of different small groups than strong ties and are useful in transferring new knowledge. Strong ties are useful in transferring complex knowledge. Strong ties provide a base of trust that can reduce resistance to change. Successful transition is conditional on the linkage of otherwise disconnected social networks. This is clearly shown in this case study, as successful adoption of hybrid seed was conditional on new trading and information channels.

In this case study, social networks play a role particularly in the acceleration phase of the transition and not so much in the pre-development phase. During the pre-development phase, in which seed companies were experimenting, there was little interaction between actors and systems as described in Luhmann's work. Farmers' adoption of newly developed varieties does require social networks. These networks relate to the ties among farmers and those between farmers and their suppliers and buyers. The ties between seed companies, traders and farmers are assumed to be weak. These weak ties did transfer new knowledge, namely the existence of improved vegetable seed. The tacit knowledge of this seed (like its characteristics and cultivation requirements) is unlikely to have been shared through these weak ties. This information is quite complex and its diffusion requires trust. Farmers are not quick to trust seed companies and traders, as they have a commercial interest. Farmers with strong ties do trust one another.

Social learning took place among farmers through imitation and sharing of experiences. The innovators among the farmers adopted the innovation based on information they received, probably through weak ties. However, adoption of the innovation among larger groups of farmers surely resulted from watching how the neighbour was doing, and sometimes through sharing experiences and imitation. According to Granovetter, weak ties are crucial in social learning. However, farmers have different networks, which leads to diffusion of innovations. Imitation occurs only of people who are trusted to be doing well.

10.4.4 Doubt management: Weick

Weick's theory relates to the limited capacity of individuals and organisations to learn and deal with future uncertainties. According to Weick, people face two challenges when they create maps of the future. Either they become overconfident or they risk becoming overly cautious. In the process of making sense of things, people engaged in change often feel obliged to use a rational order and clear goals as feedstock for transition. Weick emphasises the limited rational capacity of human beings. The key for Weick is to establish tolerance for ups and downs in the sense of social order by accepting those as normal fluctuations rather than perceiving them as some kind of testimonial to one's own shortcomings.

In this case, the fix of the system is seen in breeding vegetables and, indirectly, by following a BoP organisation approach. The seed company had no doubt that the local information needed for the breeding process would be available to them. The company was confident in its belief that improved seed materials was a rational solution to improve smallholder farmers' income (as well as to its own earnings). Being a profit-oriented enterprise, the seed company tried to avoid high levels of uncertainty in marketing the seed. Its strategy was to look for local companies to join in the marketing, rather than organising all of the marketing itself.

An interesting lesson from this business case is that the behaviour of companies tends to be rather straightforward starting from practice. It can be assumed that in such an incremental process, sensemaking is done in relation to actual problem solving. Weick argues that all of these actions are cryptic, incomplete and tentative. But what may happen when companies become part of strategies that address complex problem, the solving of which may require connectivity with a wide range of different actors? Weick then warns of the risk of simplifying matters, leading to a situation wherein accuracy may cause inertia.

The company's strong focus on developing improved plant varieties for specific contexts indicates that in this change process people concentrated on a few critical problems, learnt the history of them, built coalitions, and mobilised support. Weick might conceptualise this business-driven change in terms of a loosely coupled organisation that remains rather stable by its flexible learning ability. Organisations can produce continuous change by applying improvisation, translation and learning, which seems to contrast with a presupposed and orchestrated plan. But it also implies that such organisations are deeply embedded in hands-on problem solving and do not have enough distance from the original problem to really orchestrate a change process in, for example, the terms under which business is performed.

10.4.5 Power and discipline: Foucault

According to Foucault, knowledge, truth and rationality are social constructions which are produced in social systems. Knowledge is used as a powerful tool to reach goals. It is possible to decide which knowledge enters a system and which does not. Therefore, systems create their own reality. It is important not to neglect unintended effects. In the case study at hand, the power of knowledge is with the private sector. However this only accounts for the knowledge on how to produce the hybrid, which is costly and exclusive knowledge. The knowledge on how to grow the seed and the additional techniques required is also necessary in order to have a successful crop. The success of the diffusion of hybrid varieties depends on the existence of this knowledge and farmers' experience in agriculture.

Detecting the mechanisms of disciplining, subjection and exclusion can be of help in the analysis of the introduction of hybrid planting material to smallholder vegetable farmers. These mechanisms often remain hidden, are fired unconsciously and have unintended consequences. This case shows the discourse of seed companies as assuming smallholder farmers are market oriented and therefore willing to change and adopt hybrid seed varieties. Hence, selling seeds, which seems to be a straightforward act, may turn farmers into certain subjects and also discipline their economic behaviour and social choice. This discourse can exclude farmers who are not market oriented or who decide to cope with risks differently. Exclusion further occurs because not all farmers have access to hybrid varieties. This is the case, for example, when farmers lack the required starting capital or the appropriate marketing channels.

10.4.6 Materiality: Benton

The materialist perspective of Benton contributes to our understanding of the changes in socio-technical regimes taking place in vegetable production in South-East Asia. The transition literature maintains a strong dichotomy between the material-biological and the social-human, while NIE devotes little attention to the nature (in both senses of the word) of technology. In this case, the technology concerns agriculture, which Benton argues should be seen as a labour process in which natural processes are regulated, unlike the industrial labour process in which 'raw' materials are transformed.

Regarding newly bred vegetable varieties, Benton's perspective suggests that these varieties can be interpreted as the product of earlier efforts to regulate biological processes. The associated cultivation technology involves the use of chemicals, fertilisers and pesticides, as well as physical structures such as trellises and raised beds, all to regulate the plant environment. The technology aims to transcend naturally imposed limits on the productivity of existing plant varieties following a paradigm of value maximisation.

A defining characteristic of this new socio-technical regime is the change in the scope of the regulatory nature of the agricultural labour process. The hybridisation technology engenders a transfer of the activities of selection and their switching from farmer to breeder. The farmer relinquishes this activity, and substitutes it with a commercial transaction, with the expectation of transcending the limits of the natural environment. The new social relation between farmer and seed provider entails a shift in dependency relations. Within local farmer-derived seed systems, the farmer was dependent to varying degrees upon others in the immediate environment for the availability of seed. With the commercial relation, this dependency is raised in degree, but also shifted to new foreign actors, represented by the seed company.

The success (in terms of persistence) and evolution of this socio-technical regime depends, however, on natural factors that still impose limits on the vegetable production system. In the case study, this was clearly visible in a high variability in profitability across farmers that accompanied the overall average increase. In relatively simple terms, the new 'transcending' technology can be viewed as displacing a 'traditional' technology characterised more by a principle of adaptability to natural processes.

Another unintended consequence might be genetic erosion – loss of genetic diversity – among vegetable crop species. This could potentially constrain the regulatory labour efforts of breeders in the future. But the extent of this process was not examined in the case study. Nor did the case study assess other potential unintended consequences in relation to health and environment. In general, though, the diffusion of the transcending technology follows the same logic of intensive agricultural production as that pursued in industrialised countries, which is recognised as not having foreseen many externalities. For example, industrialised countries have met formidable challenges in regulating farmers' practices, but these are still less daunting than those in South-East Asia. In rice production in South-East Asian, the institutional context (and underlying social relations) made it even more difficult to place human limits on the ecological and health damage incurred.

10.5 Conclusions

This chapter applied a selection of theoretical arguments to the case of a business-driven way of doing things differently. Doing so opened the black box of what seemed to be a purely practical, business-driven and problem-oriented process. It may also have produced new types of questions. For example, what would happen if the relatively closed business environment were to link with change processes in the public domain aimed to improve property rights or market regulation? Or, what would happen if a wider group of stakeholders became involved in order to construct a creative coalition and therefore build a system which is more open to new ideas and wherein a new balance between weak and strong ties evolves? The exercise in this chapter suggests that to explain the adoption of hybrid varieties it may be insufficient to focus only on the properties and technical effects of the seeds themselves. Mirroring the specific business model with a wider set of social theories helped to detect other mechanisms and the institutional dynamics hidden in a business model that may explain the rate and speed of adoption.

References

Aoki, M. (2001) Toward a Comparative Institutional Analysis. MIT Press, Cambridge, MA, USA.

Breschi, S. and F. Malerba (1997) Sectoral innovation systems: technological regimes, Schumpeterian dynamics, and spatial boundaries. In: C. Edquist (ed.). Systems of Innovation: technologies, institutions and organizations. Pinter, London and Washington, pp. 130-156.

Danse, M.G. and S. Vellema (2007) Small scale farmer access to international agri-food chains: a BoP-based reflection on the need for socially embedded innovation in the coffee and flower sector. Greener Management International. June 2007 issue: 39-52. Available at: http://goliath.ecnext.com/coms2/gi_0199-6701036/Small-scale-farmer-access-to.htm.

EWS (East West Seed) (2002) Vegetable breeding for marketing development. East West Seed International, Nonthaburi, Thailand.

Eaton, D. and R.C. Wiersinga (2009) Impact of Improved Vegetable Farming Technology on Farmers' Livelihoods in Asia. LEI Wageningen UR, Den Haag, the Netherlands.

Geels, F.W. (2004) From sectoral systems of innovation to socio-technical systems: insights about dynamics and change from sociology and institutional theory. Research Policy 33: 897-920.

Geels, F. and R. Raven (2006) Non-linearity and expectations in niche-development trajectories: ups and downs in Dutch biogas development (1973-2003). Technology Analysis and Strategic Management 18 (3/4): 375-92.

Hart, S.L. and S. Sharma (2004) Engaging fringe stakeholders for competitive imagination. Academy of Management Executive 18 (1): 7-18.

Langlois, R.N. (1992) Transaction-cost economics in real time. Industrial and Corporate Change 11 (1): 99-127.

Latour, B. (1992) Where are the missing masses? The sociology of a few mundane artefacts. In: W.E. Bijker and J. Law (eds.). Shaping Technology/Building Society. MIT Press, Cambridge/London, pp. 205-224.

London, T. and S.L. Hart (2004) Reinventing strategies for emerging markets: beyond the transnational model. Journal of International Business Studies 35: 350-70.

Prahalad, C.K. (2004) Fortune at the Bottom of the Pyramid: eradicating poverty through profits. Wharton School Publishing, Upper Saddle River, NJ, USA.

Rotmans, J., D. Loorbach and R. Van der Brugge (2005) Transitiemanagement en duurzame ontwikkeling: co-evolutionaire sturing in het licht van complexiteit. Beleidswetenschap 19 (2): 3-23.

Teece, D.J., G. Pisano and A. Shuen (2001) Dynamic capabilities and strategic management. In: G. Dosi, R.R. Nelson and S. Winter (eds.). The Nature and Dynamics of Organizational Capabilities. Oxford University Press, Oxford, UK, pp. 334-62.

Winter, S.G. (1984) Schumpeterian competition in alternative technological regimes. Journal of Economic Behavior and Organization 5 (3-4) September: 287-320.

Chapter 11

Changing the crop protection or pesticide use regime in the Netherlands: an analysis of public debate

Jan Buurma

11.1 Introduction

From 1998 to 2007, a fierce public debate raged on crop protection and pesticide use in the Netherlands. That debate dramatically changed the relationship between the agricultural community on one side and the general public on the other side. The debate was not managed by the government or by the agricultural community. Rather, it was the other way round: a social movement got under way against the reluctance of the government to protect the general public and against the environment hazards and risks of pesticides. At the regime level, the public debate resulted in forms of compliance with maximum residue limits (MRLs) on produce and environmental criteria for pesticide registration.

This chapter examines the public debate on crop protection and pesticide use in terms of frequencies of questions raised in parliament and articles published in national newspapers. This descriptive analysis shows the relative importance of distinct themes and changes in these themes over time. It also demonstrates which public and private organisations took the lead in the public debate on the various subjects. The descriptive analysis is followed by a dramaturgic analysis in which the decisive moments and actors for change are identified. Furthermore, reflections are presented on how the change in socio-technical regime could happen, where the seeds of transition were sown, which actors and stakeholders disturbed the established order and which actions were decisive for the regime changes. In the final section, a selection of theoretical frameworks is applied to deepen the analysis and to detect candidate mechanisms for explaining how this change of socio-technical regime came about.

11.2 Reconstruction of the public debate on crop protection and pesticides

A total of 279 documents (parliamentary questions and national newspaper articles) on crop protection and pesticides were retrieved from specific websites (www.overheid.nl and http://lexisnexis.academic.nl). These documents were classified in ten themes based on their subject matter. The major themes (in frequencies of documents) in the public debate were registration policy in conjunction with stakeholder relations and food safety in conjunction with labour safety and public health. The major actor groups in the debate on pesticide registration were non-governmental organisations (NGOs) and agricultural sector organisations. The major actor groups in the debate on food safety and labour safety were NGOs and members of parliament. Table 11.1 presents document statistics reflecting the development of the public debate on crop protection and pesticides in the Netherlands.

A review of the public debate, furthermore, points to a series of minor themes, such as invasive species, genetic modification, integrated production, inspection and certification, water pollution and competitive position. These minor themes were dominated by the major themes in the 1998-2006 period. Before 1998, genetic modification, water pollution and inspection were important subjects. After 2006, invasive species, water pollution, competitive position and genetic modification gained more importance. This pattern indicates a structural change in the content of the debate in 1998 and again in 2007. Simultaneously the actor groups changed as well. Before 1998 and after 2006 questions in parliament dominated the debate. Between 1998 and 2006, articles in national newspapers carried the debate.

Analysis of the documents reveals a striking difference in word choice with regard to pesticides and crop protection between the various actor groups in the debate (Table 11.2).

Table 11.1. Numbers of documents (parliamentary questions and national newspapers) classified in themes, years and actor groups.

Theme	1995	1996	1997	1998	1999	2000	2001	2002	2003	2004	2005	2006	2007	2008	Total
Invasive species	1	2	0	1	0	1	1	0	0	1	0	2	4	4	17
Genetic modification	2	3	1	2	1	3	0	0	2	1	1	0	0	2	18
Integrated production	0	0	1	1	0	1	1	2	2	3	3	2	0	1	17
Health effects	2	4	1	6	2	0	0	0	3	1	1	0	1	2	23
Food safety	0	0	0	0	2	9	4	3	4	14	6	2	3	3	50
Registration policy	0	6	1	3	17	8	14	6	12	7	4	5	5	1	89
Stakeholder relations	0	2	0	0	3	1	0	3	7	3	5	0	0	0	24
Inspection&certification	1	0	2	2	1	3	0	0	4	2	0	1	0	0	16
Water pollution	3	2	0	2	2	1	0	0	0	0	1	2	0	3	16
Competitive position	1	1	0	0	1	0	1	0	0	2	0	0	2	1	9
Total of documents	10	20	6	17	29	27	21	14	34	34	21	14	15	17	279
Colour of Government	social-liberal				social-liberal				centre-social				centre-social		
Minister of Agriculture	Van Aartsen				Brinkhorst				Veerman				Verburg		

Parliament		Parliament + sector orgs		Sector orgs + universities	
Parliament + universities		Parliament + NGOs		Sector orgs + NGOs	

Table 11.2. Quantitative presentation of word choice: 'pesticides' or 'crop protection'.

	Reference = pesticides		Reference = crop protection	
	number	percentage	number	percentage
Parliamentary questions				
Ministry of Agriculture	16	37%	27	63%
Other ministries	26	76%	8	24%
Parliamentary questions				
Left wing parties	30	77%	9	23%
Central parties	15	56%	12	44%
Right wing parties	4	36%	7	64%
Newspaper articles				
Agricultural Daily	62	61%	39	39%
General newspapers	98	97%	3	3%
Newspaper articles				
Sector organisation	53	63%	31	37%
Non-govt organisations	81	91%	8	9%
Universities	26	90%	3	10%

Table 11.2 shows a preference for the word 'pesticides' among the actor groups representing the general public, whereas actor groups representing the agricultural community preferred speaking of 'crop protection'. Apparently the general public wanted to emphasise the negative aspects of agro-chemicals (pesticides), whereas the agricultural community wanted to emphasise the positive aspects (crop protection). Figure 11.1 summarises the composition of the actor groups shown in Table 11.2 as representing the general public and the agricultural community.

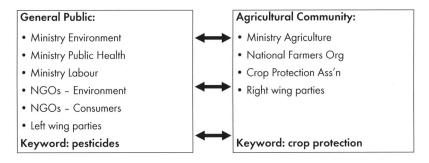

Figure 11.1. Characterisation of two sides of the public debate.

11.2.1 The debate on food safety

The debate on food safety in conjunction with labour health and safety started in 1995 with questions posed by left-wing members of parliament on the effects of pesticides on public health (particularly in relation to brain tumours, cancer and reproductive health). The Minister of Public Health responded that scientific evidence of a causal relationship was lacking. In 1998 a consortium of NGOs launched a public-awareness campaign on pesticide residues in lettuce. They compared a head of lettuce with a condom: two ways to prevent conception. They asked the government to immediately ban seven pesticides frequently used in vegetable production. The government again stated that scientific evidence was lacking. They accused the NGOs of inciting panic among the general public.

The NGOs concluded that the government was unresponsive to their questions and claims. For that reason they shifted the focus of their campaigns to supermarkets. In 2000 they bought boxes of strawberries from supermarkets and had a well-respected laboratory analyse them for pesticide residues. Many unauthorised pesticides were found, as well as amounts of chemicals exceeding maximum residue limits (MRLs). They filed a complaint against the supermarkets with the public prosecutor in Amsterdam. The supermarkets responded that the strawberry growers were responsible for the residues. The growers, in turn, pointed to strawberry growers in Belgium, who had access to pesticides that were banned in the Netherlands.

Some months later the NGOs requested the Minister of Public Health to immediately stop the sale of vegetables and fruits with agro-chemical residues exceeding the allowed levels. They claimed that each day 50,000 Dutch children ingested too much pesticide, leading to behavioural disorders. The Minister replied again that scientific evidence was lacking and that the conclusions of the NGOs were disputable. A left-wing member of parliament posed a question to the Minister of Public Health on the risks to the public health of exceeding MRLs in vegetables and fruits. The Minister answered that the MRLs were set much lower than acceptable daily intakes. For that reason she denied that exceeding the MRLs posed a threat to public health. She also said that the Food Inspection Authority monitored the MRLs.

In 2002, the NGOs started a campaign on pesticide residues in grapes from Greece and Italy. They took two supermarket concerns to court because produce on their shelves had residues exceeding the MRLs. The two supermarkets reached an agreement with the NGOs on compliance with the MRLs. A left-wing member of parliament confronted the Minister of Public Health with the findings on pesticide levels in grapes, raising doubts about the performance of the Food Inspection Authority. The Minister said that the high residue levels found on the grapes was caused by a difference between countries in authorised pesticide applications.

In 2003, NGOs started a campaign to raise awareness of pesticide residues in nectarines and grapes. The produce of some supermarkets was said not to be in compliance with the MRLs. The Food Inspection Authority became angry and accused the NGOs of spreading misleading information and consequently making consumers afraid of vegetables and fruits.

In 2004 the NGOs organised a demonstration of blindfolded consumers at the Ministry of Agriculture. They asked the Minister to disclose the Food Inspection Authority's MRL data. The Minister refused. The NGOs lodged an appeal against the Ministry and finally got the MRL data from the Food Inspection Authority. In the data the NGOs found examples of unauthorised pesticide use. Publication of the data caused much irritation among growers, supermarkets and politicians. The Central Bureau of Food Retailers requested that the government take measures against the NGOs.

In 2005 the NGOs published a monitoring report on pesticide residues in lettuce. They found just one sample with residues exceeding the maximum. Within days that supermarket and its supply company traced the grower who had produced the lettuce. In 2007 the Food Inspection Authority published a report showing a decreasing trend in produce with residues exceeding the MRLs. NGOs acknowledged the positive trend, but still found MRL problems in grapes in the data of the Food Inspection Authority. The Central Bureau of Food Retailers again accused the NGOs of spreading misleading information.

In 2007 the NGOs moved their campaigns to the European level. They bought fruit samples in the supermarket at the European Parliament. Laboratory tests found residue limits exceeding the maximum in three of the eight samples. The NGOs used this finding to put pressure on EU parliamentarians to advocate a strict Framework Directive limiting authorised pesticide use at the EU level. The Dutch National Farmers Organisation, however, feared the restrictions proposed in such a Framework Directive and defended the current, science-based, authorisation procedures.

The timeline of the NGO campaigns on public health and food safety shows that the government systematically rejected the claims of the NGOs. They repeatedly said that scientific evidence was lacking. The supermarket concerns were much more susceptible to the NGOs' message. While they accused the NGOs of spreading misleading information, they were nonetheless very much afraid of losing consumer confidence. For that reason, the supermarkets forced supplier and growers to comply with the MRLs. In fact the NGOs managed to change the socio-technical regime of compliance with MRL requirements. The supermarkets effectively took over the responsibility for ensuring the MRLs from the Ministry of Agriculture.

11.2.2 The debate on pesticide registration

The debate on pesticide registration started in 1996 after publication of the mid-term evaluation of the Multi-Year Crop Protection Plan. A consortium of seven NGOs asked the Minister of Agriculture to ban a series of pesticides, to introduce field margins and to impose taxes on pesticide use. The Minister refused because of bottlenecks in pest control for minor crops and fears that national prohibitive measures would just stimulate the import of pesticides from neighbouring countries.

In 1998 the NGO *Stichting Natuur en Milieu* (SNM) started a lawsuit against the Ministry of Agriculture. SNM stated that the authorisation of the fungicide chlorthalonil was at odds with the regulations. The Minister explained in a letter to parliament, that the authorisation had been the result of an agreement (made in 1993) between government, the chemical industry and farmers on flexibility in authorisation of chlorthalonil in the 1995-2000 period.

In 1999 a centre party parliamentarian asked the State Secretary of Agriculture about the consequences of the withdrawal of authorisation for 40-50 active ingredients as of 1 January 2000. The State Secretary answered that the withdrawal was a result of the administrative agreement (made in 1993) with farmers and the chemical industry regarding a clean-up operation of the pesticide package. Three months later, SNM again started a lawsuit against the Ministry of Agriculture. SNM claimed that the Board for the Authorisation of Pesticides had re-authorised several pesticides without checking their impact on the environment. Even pesticides known to have harmful effects had been re-authorised. The Ministry defended the re-authorisation of dichlorvos because of its 'essentiality' for integrated pest management in greenhouses.

In October 1999 SNM and the National Farmers Organisation (LTO) initiated explorative talks on opportunities for a deal on the withdrawal of 42 pesticides as of 1 January 2000. The chairperson of the Executive Board of the Multi-Year Crop Protection Plan (a politician) suggested banning the most harmful pesticides and saving the most 'essential' applications of pesticides. The government established a committee (including NGOs and LTO) to learn which of the 42 pesticides could be labelled 'agro-technically essential'. The two parties agreed on 90% of the applications, but the last 10% formed a stumbling block. The committee exploded.

In 2000 SNM started a lawsuit against the Ministry of Environment. It accused the government of tolerating several 'essential' pesticides in the market, without a legal basis for doing so. The Minister of Environment had no choice but to ban the 'essential' pesticides. Growers of vegetables and fruits were desperate. The evaluation process of 'essential' pesticides was accelerated. In 2001, however, the chemical industry failed to deliver complete dossiers for the 'essential' pesticides. Consequently the 'essential' pesticides were banned. The farmers were angry and frustrated. They discontinued their cooperation in the pilot project A View to Crop Health, which was designed to validate the targets of the second National Action Plan.

In 2002, SNM started a lawsuit against the Board for the Authorisation of Pesticides. The Board wanted to give provisional authorisations for maneb and mancozeb in anticipation of European authorisation. According to SNM this procedure was against existing legislation. The court refused to allow the provisional authorisations. The chemical industry, however, was only interested in producing dossiers for the European authorisation process. Consequently 150 pesticides fell by the wayside, and the Dutch registration policy ended up in an impasse.

In August 2002 the new Minister of Agriculture gave an exemption for Ridomil, because onion growers were in urgent need of fungicides to save their crop from mildew. SNM announced that it would not dispute the exemption. In February 2003 the Ministry of Agriculture together with the Ministry of Environment, LTO en SNM agreed on 32 authorisations for minor crops. The impasse had been cleared.

In March 2003 the Minister of Agriculture forged a covenant with farmers, the chemical industry, SNM, water companies and the water boards. The target of the covenant was a 95% reduction of the environmental impact of pesticides. In October 2003 the National Farmers Organisation (LTO) presented the sector plans formulated in the framework of the covenant. LTO chairperson stressed that the sector needed a wide package of pesticides to meet the target of 95% environmental impact reduction. SNM was not amused. Provincial branches of SNM started action against the covenant. Some weeks later SNM withdrew from the covenant.

In 2004 the Ministry of Agriculture announced exemptions related to 40 agro-technical bottlenecks. The Ministry wanted to solve the bottlenecks in minor crops (in anticipation of authorisation at the European level). A provincial branch of SNM started several lawsuits against the Ministry. The court ruled that there was no legal basis for the exemptions. The authorisations were undone. Farmers sounded the alarm. The Minister announced the development of a new pesticide law.

In 2007 the Ministry of Agriculture presented its new crop protection law, which was attuned to European policy. The Board for the Authorisation of Plant Protection Products and Biocides was no longer an instrument of policy-makers and politicians in The Hague. The Dutch Crop Protection Association, the National Farmers Organisation and the NGOs were happy with the new law, which came into force 17 October 2007.

The timeline of the public debate on pesticide registration started with the remainders of an agricultural fortress (in the 1990s) in which public-private agreements were made without much democratic control. The fortress was dismantled through a series of lawsuits by SNM against the Ministry of Agriculture. The lawsuits resulted in a new institutional arrangement for compliance with the environmental criteria in the authorisation of pesticides. The concept of 'economic essentiality' brought transparency in the trade-off between the interests of the general public (environmental impact) and those of the agricultural community (yield security). On the other hand the public debate put an end to a pesticide registration policy in the Netherlands which ran ahead of EU legislation. The new pesticide law was attuned to the European policy.

11.3 The regime change in terms of transition management

This section considers the way the public debate reflects transition processes. The objective is to find lessons for transition processes in the future. Special attention is paid to the historic context of the public debate, the change in conditions feeding the public debate, the changes achieved in socio-technical regimes, the actors who tried to control the transition and the moves that were decisive for the transition's sustainability.

11.3.1 Concurrent developments

The public debate started in the 1990s, which was a turbulent period in the evolution of the Dutch agricultural sector. This was a period in which agricultural markets changed from producer-oriented to consumer-oriented. The idea of chain reversal, i.e. placing consumers rather than producers in the lead, introduced a new paradigm for Dutch farmers. Owing to overproduction, citizens and consumers could now pick and choose, leaving producers to discover what types of products they could sell in the market. The turning point in this process was the research report *Missed the Market* by Kearney (1994).

The agricultural sector also lost power in a socio-political respect. Environmental and consumer interest groups became influential. The Agricultural Board (the symbol of closed ranks in agriculture) was abolished. Agricultural research and extension were no longer free of charge. In other words, the agricultural sector was exposed to civil society and free market forces, and the collectivity of the agricultural sector was replaced by the own responsibility of farmers and growers (agricultural entrepreneurs). In summary, the power of the agricultural sector was already on the decline when the debate on crop protection and pesticides started. The power of consumers and citizens was on the rise.

11.3.2 Seeds of transition

An important milestone for the NGOs was their recognition (in 1995) as representatives of collective interests in lawsuits against government decisions. This recognition brought them in legal balance with the chemical industry (interview with Hans Muilerman, SNM). The NGOs fully exploited this recognition in initiating proceedings against the Ministry of Agriculture on the registration of pesticides. In the debate on food safety, the success of Greenpeace on the Brent Spar in 1995 probably played an important role. It became clear that a consumer strike was an important tool to put commercial companies under pressure. *Milieudefensie* (Friends of the Earth, Netherlands) fully exploited this tool to compel supermarket concerns to comply with MRLs.

11.3.3 New arrangements

The lawsuits against the Ministry of Agriculture provoked a huge public debate on economic essentiality, especially with regard to minor crops. The public debate brought transparency to the trade-off between the interests of the general public (environmental impact) and the interests of the agricultural community (yield security). A special procedure to judge economic essentiality came into force and was embedded in the European regulations governing the marketing of plant protection products. Simultaneously the centre of policy-making on pesticide registration moved from The Hague to Brussels and Dutch regulations no longer ran ahead of EU legislation.

The NGO campaigns on food safety and compliance with MRLs also brought about a change from policy control to societal pressure. The government systematically refused the claims of the NGOs that food safety was at issue. The supermarkets were much more susceptible to the claims of the NGOs. They were very much afraid of losing consumer confidence. For that reason the supermarkets demanded that their suppliers and growers comply with the MRLs. Implicitly the NGOs managed to change the socio-technical regime of compliance with MRLs. The supermarkets took over this responsibility from the Ministry of Agriculture. The NGOs kept the supermarkets under pressure by publishing the MRL analysis data of the Food Inspection Authority.

11.3.4 Leading actors controlling the transition process

The leading actor in the public debate on food safety (compliance with MRLs) was a consortium of NGOs, headed by *Milieudefensie*. They organised successive campaigns on pesticide residues in lettuce, strawberries, grapes, nectarines and other produce. The supermarkets became so fed up with the campaigns and the defensive attitude of the government, that they decided to enforce compliance with the MRLs themselves in cooperation with their suppliers in vegetables and fruits. The supermarkets were challenged by the NGOs to stick to MRL compliance by the publication of the Food Inspection Authority data on the Internet.

The leading actor in the public debate on pesticide registration (compliance with legislation on environmental criteria) was the environmental NGO SNM (*Stichting Natuur en Milieu*), supported by other NGOs. It organised the lawsuits against the Ministry of Agriculture. It forced the Ministry of Agriculture and the Ministry of Environment to develop a procedure for evaluating the economic essentiality of specific pesticide treatments. The Dutch Crop Protection Association (Nefyto) became so fed up with the lawsuits and the increasing costs of evaluation dossiers, that it decided to confine itself to dossiers for European authorisation.

This position forced the government to abandon its policy of running ahead of EU legislation and to design a new pesticide law after the European policy.

11.3.5 Conditions for the transition process

The changes described could take place only because of the declining power of the agricultural sector. This decline fits in a landscape of overproduction in agriculture, where consumers have the power to pick and choose in the marketplace, and it is up to producers to discover what types of products they can sell. The recognition of NGOs as representatives of collective interests in lawsuits against government decisions produced a strategic niche for the NGOs to undermine the agricultural fortress (socio-technical regime) in which public-private agreements were made without much democratic control by citizens and consumers.

Disturbance of the existing socio-technical regime by the NGO campaigns and lawsuits forced input suppliers and supply chain partners in the agricultural community to take responsibility or to take action. They saw themselves confronted with economic hazards emanating from consumer strikes and administrative burdens. A remarkable phenomena was the defensive attitude of the government in this case. Government continuously hid behind the lack of scientific evidence, the lack of proven causal relationships and misleading information. In other words, the public sector was inclined to endorse the existing socio-technical regimes.

In conclusion, the public debate on crop protection and pesticides and the resulting transition to a more socially responsible crop protection regime was the result of changes at both the macro level (the socio-economic landscape) and the meso level (the socio-technical regime) and new approaches at the micro level (societal novelties). Figure 11.2 depicts these changes and novelties.

The changes at the macro level are the chain reversal (marked by the 1994 report *Missed the Market*) and the rise of consumer power (marked by the Brent Spar action of 1995). The changes at the meso level are recognition of NGOs in lawsuits against government decisions (1995) and abolition of the Agricultural Board (1996). The new approaches at the micro level included the campaigns and lawsuits of the NGOs against the supermarket concerns and against the government. The campaigns and lawsuits were first organised in 1998. The dates show that the initial changes in the socio-technical regime and in the socio-economic landscape were not caused by experiments and novel approaches at the grassroots level. Rather, it was the other way around: the changes at the landscape and regime level provided a strategic niche for the NGOs' campaigns and lawsuits. This finding contrasts with the general opinion in transition literature, that change starts with technological novelties at the micro level. This analysis shows that transition processes can also start with socio-economic and socio-technical changes at the macro and meso level. Nevertheless, the multi-level perspective of transition theory is very useful for understanding the course and the outcome of the public debate on pesticides and crop protection.

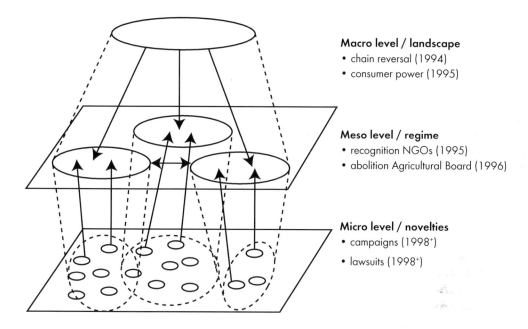

Macro level / landscape
- chain reversal (1994)
- consumer power (1995)

Meso level / regime
- recognition NGOs (1995)
- abolition Agricultural Board (1996)

Micro level / novelties
- campaigns (1998+)
- lawsuits (1998+)

Figure 11.2. A multi-level perspective of the public debate on pesticides and crop protection (adapted from Geels, 1997: 58).

11.4 Theoretical reflection

This section links the patterns discovered in the public debate (in both the general description and in the detailed analyses of the debates on food safety and registration policy) to the different social theories discussed in this book. Key elements of the different theories are mentioned and practical findings in the analysis of the public debate are indicated.

11.4.1 New institutional economics: North and Aoki

The theory of new institutional economics starts with agents which have certain beliefs about the world. Two important agents in the public debate on pesticides are the NGOs *Stichting Natuur en Milieu (SNM)* and *Milieudefensie (Friends of the Earth, Netherlands)*. According to SNM, the government has a duty to protect the environment. The focus of *Milieudefensie* is more on the role of human actions in the deterioration of the environment. This difference explains why SNM focused on lawsuits against the government (on pesticide registration) and *Milieudefensie* focused on campaigns against supermarket concerns (on compliance with MRLs).

Figure 11.1 presented the individual players in the public debate: government ministries, NGOs and political parties. These were organised in two opposing camps representing the general public and the agricultural community. The domains of the game are the environmental and the agricultural impacts of pesticides. The drastic changes in the 1990s (chain reversal, abolition of the Dutch Agricultural Board, recognition of NGOs as representing civil society and the rise of civic action reflected in the Brent Spar endeavour) gave the NGOs the behavioural belief that strategies like lawsuits and campaigns could produce a new equilibrium between the general public and the agricultural community and new institutions with regard to pesticides and crop protection. The series of lawsuits and campaigns described in the previous sections indeed resulted in new rules and symbols: compliance with MRLs enforced by supermarket concerns (instead of by government), a regulation for evaluating 'economic essentiality' and the new Law on Crop Protection and Biocides attuned to EU regulation on pesticides, including a more independent Board for the Authorisation of Plant Protection Products and Biocides.

An important issue in NIE is coordination between multiple actors with different interests and information. The multiple actors and their interests in this case are the general public (aiming at reducing the risks of pesticides for health and environment) and the agricultural community (aiming at using the blessings of plant protection products to secure agricultural yields). The conflicts of interests between the two opposing actor groups were balanced through (1) recognition of NGOs in lawsuits against government decisions, (2) environmental criteria being set for the registration of pesticides and (3) establishment of a procedure for evaluating economic essentiality.

The 'entrepreneurs' (individuals who induce change) in the public debate on pesticides were the NGOs. Through their campaigns and lawsuits they caused institutional crises in the fields of food safety and pesticide registration. In the case of food safety, the supermarkets enforced compliance with MRLs through enforcing Global Good Agricultural Practice (Global GAP). In the case of pesticide registration, policy-makers introduced the concept of 'economic essentiality'. Later on, the chemical industry enforced the adjustment of dossier requirements to forthcoming EU regulations.

11.4.2 Social systems: Luhmann

Luhmann conceptualises social systems as self-reproducing networks of communication. Important concepts in his theory are autopoiesis, code of communication, lack of observation and path dependencies. All of these concepts are present in the public debate on food safety. Evidence of self-reproducing networks of communication was given in Table 11.2 and in Figure 11.1, which showed that representatives of the general public prefer the word 'pesticides' and representatives of the agricultural community prefer speaking of 'crop protection'. In fact, these two wordings are codes of communication of the two opposing networks.

A banal example of autopoiesis in the public debate on pesticides and crop protection is the selection of wordings in parliamentary questions: a left-wing member of parliament submits a question on 'pesticides' to the Minister of Agriculture, and the Minister responds using terms like 'plant protection products' and 'crop protection'. A more serious example is the Babel-like confusion on food safety. NGOs, for strategic reasons, associate food safety risks with foodstuffs exceeding the MRLs. The government, however, systematically rejected this association, repeatedly explaining that the Acceptable Daily Intake (ADI) and Acute Reference Dose (ARfD) are in fact the proper indicators of food safety. As a result there was hardly any discussion of the fact that exceeding MRLs is against the law (i.e. legal instructions for use).

This brings us to the limits of governmental monitoring systems. Government agencies were unaware that exceeding the MRLs implied a potential risk for supermarkets. Breaching the confidence of the general public in the safety of vegetables and fruits on supermarket shelves is catastrophic for a store's reputation and for the economic margins of the parent concern. This explains why the supermarkets took over the enforcement of compliance with MRLs from the government. The operational closure of the government (its strict focus on the ADI and ARfD) prevented it from understanding the impact of the NGO campaigns on the supermarkets.

11.4.3 Strong and weak ties: Granovetter

The social networks theory of Granovetter centres on the relevance of strong and weak ties for information exchange between actor groups and for the diffusion of innovations. The opposing actor networks presented in Figure 11.2 reveal strong ties within and weak ties between the two opposing actor groups. In particular, the champions of pesticide risk reduction (the left block) stick together and the champions of agriculture production innovation (right block) stick together. There is little interaction (weak ties) between the two opposing blocks. According to Granovetter, weak ties between the two opposing actor groups play a crucial role in social learning.

The NGOs played an important role in crossing the *no man's land* between the general public and the agricultural community. They went to the Ministry of Agriculture (through lawsuits) and to the supermarket concerns (through campaigns). As a result, the public's worries about pesticide residues in the environment got through to the agenda of the Ministry of Agriculture. As a result, the environmental criteria for the registration of pesticides were taken more seriously and a formal procedure was established for assessing economic essentiality. The chemical industry also played an important role in crossing the *no man's land* between the national pesticide registration policy in the Netherlands and the harmonised pesticide registration policy in Europe.

Simultaneously public concern about pesticide residues in vegetables and fruits got through to the agendas of the supermarkets, suppliers and primary producers. As a result the supermarkets enforced practically full compliance with MRLs through broad implementation of Global GAP in their supply chains for vegetables and fruits.

The aggressive campaigns of the NGOs resulted in transition arenas in which both private and public partners were involved. The first example was the explorative talks of the NGO *Stichting Natuur en Milieu* and the National Farmers Organisation *LTO* in October 1999 to reach a deal on the withdrawal of 42 pesticides as of 1 Janaury 2000. The second example was the covenant on sustainable crop protection, established in March 2003. In the covenant the Ministry of Agriculture, the Ministry of Environment, farmers, the chemical industry, *Stichting Natuur en Milieu*, water companies and water boards joined forces to realise a 95% reduction of the environmental impact of pesticides by 2010.

11.4.4 Doubt management: Weick

Weick explains that a lack of sound doubt blocks out any progress of thought. A relentless fixation on one's own convictions prevents people from finding solutions to organisational problems. For that reason, organisations need loosely coupled elements, i.e. a flexible learning ability. Sound doubt was observed at several places and points in time during the public debate on pesticides. The most obvious examples are presented here.

In 2002 two supermarket concerns reached an agreement with NGOs on compliance with MRLs. This was quite surprising, because in the years before and after, the Central Bureau of Food Retailers accused the NGOs of spreading misleading information. The supermarkets had the option of following the government's example in explaining to consumers that scientific evidence was lacking. Apparently they were not confident that consumers would accept the scientific argument.

In October 1999 *Stichting Natuur en Milieu* started explorative talks with the National Farmers Organisation to look for opportunities for a deal on the withdrawal of 42 pesticides as of 1 January 2000. They also accepted the government's invitation to join the committee to sort out which of the 42 pesticides could be labelled as 'agro-technically essential'. This was an experiment, because in the years before and after they started several lawsuits against the government in order to enforce the withdrawal of the 42 pesticides. Apparently they were not confident that the best results could be achieved through lawsuits.

In 2002 the chemical industry refused to provide the dossiers for a national re-evaluation of pesticides. Rather, they wanted to enforce a switch-over to the forthcoming European authorisation regulations. Apparently they were not confident that meeting the Dutch dossier requirements would stop the NGOs' lawsuits against the Board on the authorisation of pesticides.

Lack of sound doubt is observed in the behaviour of the Ministry of Agriculture and the Ministry of Public Health. They systematically rejected the claims of the NGOs that failure to comply with MRLs was a danger to public health. Owing to their fixation on Acceptable Daily Intakes and Acute Reference Doses they forgot to address the overruns of MRLs in vegetables and fruits. This defensive attitude led the supermarket concerns to decide to tackle the MRL problems directly with their supply chain partners.

The transition to private sector supervision on compliance with MRL requirements is a striking example of episodic change under the pressure of consumer strikes. Another example in the public debate on pesticides is the regulation on economic essentiality. This regulation arose under pressure of the withdrawal of a whole bunch of pesticide applications.

11.4.5 Power and discipline: Foucault

Discourse is a basic element in the conceptual framework of Foucault. The French word *discours* represents an environment of strategic games of action and reaction, question and answer, domination and evasion, as well as struggle. Important mechanisms revealed by Foucault are disciplining, subjection, exclusion and evasion. These mechanisms were evident in the public debate on crop protection and pesticides. Examples are given below.

The mechanism of disciplining was especially used in the public debate on food safety. The National Farmers Organisation defended science-based procedures for pesticide authorisation. The Central Bureau of Food Retailers emphasised that the Food Inspection Authority was the proper reference point for information on MRL compliance. By the same token, the government said that the European Food Safety Authority was the right organisation to evaluate the health risks of pesticides.

The mechanism of subjection was also evident in the public debate on food safety. Both the government and the supermarkets tried to depict the NGOs as subversive organisations, who were set on inciting panic among the general public, whose conclusions were disputable, who were spreading misleading information and making consumers afraid, and who were acting against the agency that according to the government was responsible for taking measures if needed.

The mechanism of exclusion was frequently used in the public debate on labour safety. Both the Minister of Public Health and the Minister of Agriculture often cited the lack of a causal relationship between pesticides and labour health risks. They maintained that scientific evidence was scant and disputed the NGOs' conclusions. This was in the 1995-2000 period. From 2000 the public debate shifted to compliance with MRLs and NGOs' requests that supermarket concerns take responsibility for MRL compliance. In this period, the mechanism of exclusion was used again. Both the Food Inspection Authority and the

Central Bureau of Food Retailers accused the NGOs of spreading misleading information and making consumers afraid of vegetables and fruits.

The mechanism of evasion was used continuously in the reactions of the government to requests and lawsuits filed by *Stichting Natuur en Milieu* with regard to pesticide registration. In the first years (1995-1999) government spokespeople mentioned finding pragmatic solutions and conducting a clean-up operation. In other words, the government tried to communicate that it was working on the problem. As the deadline for the withdrawal of 42 pesticides (1 January 2000) came into sight, the government could no longer maintain that a soft landing was still possible. The concept of 'agro-technical essentiality' (later 'economic essentiality') was therefore introduced. Later in the debate (2002-2004) the words 'provisional' and 'exemptions' were frequently used. With these words the government tried to communicate that a full solution was within reach.

11.4.6 Materiality: Benton

The work of Benton is interesting in the interaction between the social and biological domains. The idea is that intentional processes of change can be hampered by natural and biological limits. This idea is confirmed in the public debate on pesticide registration. Examples are given below.

In August 2002 the new Minister of Agriculture (a farmer himself) saw onion growers struggling to control powdery mildew in their onion crops. The farmers had the choice of either accepting heavy crop losses or using non-authorised fungicides. The Minister realised that neither of these options was acceptable and gave the onion growers an exemption for the fungicide Ridomil. This was an example of a natural limit that hampered the intentional process of reducing the environmental risks of pesticides.

The concept of *economic essentiality* is another example of natural limits hampering the reduction of environmental risks. In particular, in minor crops the diversity of pests and diseases is so wide that finding environmental friendly solutions requires a long-term research effort. One could say that biological processes in minor crops are faster than innovation processes.

At the end of the day the very idea of *environmental criteria* is nature-imposed. NGOs have the conviction that soil life is the basis of the food pyramid on earth: soil organisms feed plants, plants feed animals and humans, animals provide milk and meat for humans. NGOs are afraid that 'a nation that spoils its soil, destroys itself' (to use the words of US President Franklin D. Roosevelt 1937). In this context, the environmental criteria are meant to protect soil organisms and soil quality against chemical compounds that could possibly undermine the food pyramid.

11.5 Conclusion

The placement of the historic context of the 1990s in the multi-level perspective shows how changes at the macro level (landscape) and the meso level (regime) provided a strategic niche for the NGOs' campaigns and lawsuits. The actions and procedures can be seen as socio-political novelties. These NGO activities finally induced a series of new socio-technical practices: compliance with MRL requirements, evaluation of economic essentiality and EU-proof pesticide registration.

The suggestion made in scholarly work on transition management is that change starts in one or more technological niches at the micro level. New technologies in niches can fail or succeed. In case of success, the technology results in a transformation of the existing regime or in the emergence of a new regime. In extreme cases, the technology may finally change the socio-economic landscape. The descriptive analysis of the debate on pesticide use and crop protection clarifies how changes at the landscape level can also occur autonomously and thereby create opportunities for novelties. The analysis also made clear that novelties are not necessarily technological, but can also be socio-political. In this case they were a combination of lawsuits and new forms of collaboration. The lesson here is that transitions can best be achieved with coordination between the three levels.

The analysis of the public debate on pesticides suggests that private-sector actors were more decisive than public-sector actors in changing socio-technical practices. The supermarket concerns took responsibility for compliance with MRLs; the chemical industries decided to switch over to the EU dossier requirements for the registration of pesticides. The Dutch ministries of Agriculture, Environment and Public Health persisted much longer in defending the existing socio-technical practices and even forgot to address violations of legal regulations (environmental criteria and MRLs). The lesson learned here is that a strong fixation on one's own convictions is counterproductive for solving institutional problems, i.e. for improving socio-technical regimes.

References

Kearney, A.T. (1994) Missed the market? Owing to limited market orientation perspectives are gloomy for Dutch agro sector. Report to the Netherlands Ministry of Agriculture. A.T. Kearney management consultants, Amsterdam, the Netherlands.

Roosevelt, F.D. (1937) Letter to all State Governors on a Uniform Soil Conservation Law. 26 February. Available at: http://www.presidency.ucsb.edu/ws/?pid=15373.

Websites

Ministerie van Binnenlandse Zaken en Koninkrijksrelaties: www.overheid.nl.
LexisNexis: http://lexisnexis.academic.nl.

Contributors

Raoul Beunen

Raoul Beunen is assistant professor in the field of spatial planning, governance of nature and citizen involvement in regional transition at the Land Use Planning group of Wageningen University.

E-mail: raoul.beunen@wur.nl

Jan Buurma

Jan Buurma is senior researcher in sociology of innovation at the Agricultural Economics Research Institute (LEI) of Wageningen University and Research Centre.

E-mail: jan.buurma@wur.nl

Myrtille Danse

Myrtille Danse was head of the sustainable chain and market division at the Agricultural Economics Research Institute (LEI) of Wageningen University and Research Centre and is currently executive director of the BoP Innovation Center.

E-mail: myrtille.danse@wur.nl

Guus Dix

Guus Dix studied philosophy and sociology, he currently teaches philosophy of science, and started a PhD-project on the double constitution of 'the economy' as epistemological and political object at the University of Amsterdam.

E-mail: g.dix@uva.nl

Martijn Duineveld

Martijn Duineveld is assistant professor at the Socio-spatial Analysis group at Wageningen University, with a focus on the dynamics of power, people and places.

E-mail: martijn.duineveld@wur.nl

Roel During

Roel During is senior researcher in cultural heritage and spatial planning at Alterra of Wageningen University and Research Centre.

E-mail: roel.during@wur.nl

Derek Eaton

Derek Eaton is programme officer at the economics and trade branch of the United Nations Environment Programme (UNEP).

E-mail: derek.eaton@unep.org

Carolien de Lauwere
Carolien de Lauwere is senior researcher in agricultural entrepreneurship and agricultural systems and behaviour the Agricultural Economics Research Institute (LEI) of Wageningen University and Research Centre.
E-mail: carolien.delauwere@wur.nl

Gerdien Meijerink
Gerdien Meijerink is researcher in international development and trade at the Agricultural Economics Research Institute (LEI) and PhD candidate in the Development Economics Group of Wageningen University and Research Centre.
E-mail: gerdien.meijerink@wur.nl

Machiel Reinders
Machiel Reinders is researcher in marketing strategy and consumer behaviour at the Agricultural Economics Research Institute (LEI) of Wageningen University and Research Centre.
E-mail: machiel.reinders@wur.nl

Marc Ruijs
Marc Ruijs is senior researcher in agriculture and entrepreneurship at the Agricultural Economics Research Institute (LEI) and senior researcher in cropping and business systems at Wageningen UR Greenhouse Horticulture, both at Wageningen University and Research Centre.
E-mail: marc.ruijs@wur.nl

Trond Selnes
Trond Selnes is researcher in public administration and policy at the Agricultural Economics Research Institute (LEI) of Wageningen University and Research Centre.
E-mail: trond.selnes@wur.nl

Catrien Termeer
Catrien Termeer is professor and chair of the Public Administration and Policy Group at Wageningen University, with a focus on processes of societal change, public leadership and new modes of governance.
E-mail: katrien.termeer@wur.nl

Kristof Van Assche
Kristof Van Assche is associate professor in the Planning and Community Development Program at Minnesota State University, with a particular interest in the appearance and history of space, and the forces molding space, spatial planning and design.
E-mail: kvanassche@stcloudstate.edu

Sietze Vellema

Sietze Vellema is assistant professor at the Technology and Agrarian Development group at Wageningen University and program leader/senior researcher Value Chains, Innovation & Development at the Agricultural Economics Research Institute (LEI) of Wageningen University and Research Centre.
E-mail: sietze.vellema@wur.nl

Gert Verschraegen

Gert Verschraegen is assistant professor at the Department of Sociology at Antwerp University, with a research interest in sociological theory, sociology of Europeanization, welfare state evolution and the sociology of knowledge.
E-mail: gert.verschraegen@ua.ac.be

Rolien Wiersinga

Rolien Wiersinga is supply chain researcher at the Agricultural Economics Research Institute (LEI) of Wageningen University and Research Centre.
E-mail: rolien.wiersinga@wur.nl